Shadworth Hollway Hodgson

The philosophy of Oeflection

Vol. 2

Shadworth Hollway Hodgson

The philosophy of Oeflection
Vol. 2

ISBN/EAN: 9783337236458

Printed in Europe, USA, Canada, Australia, Japan

Cover: Foto ©berggeist007 / pixelio.de

More available books at **www.hansebooks.com**

THE

PHILOSOPHY OF REFLECTION

BY

SHADWORTH H. HODGSON,

Hon. LL.D. Edin

AUTHOR OF "TIME AND SPACE," "THE THEORY OF PRACTICE," ETC.

IN THREE BOOKS.

VOL. II.

CONTAINING BOOK III. AND INDEX

Λεπτὰ δ᾽ ἄταρπος, νηλεὴς δ᾽ ἀνάγκα.
Aleman.

LONDON.
LONGMANS, GREEN, AND CO.
1878.

CONTENTS OF VOL. II.

BOOK III.

ANALYSIS OF PHENOMENA.

CHAPTER VII.

ELEMENTS, ASPECTS, AND CONDITIONS.

CHAPTER VIII.

NATURE AND HISTORY.

CHAPTER IX.

THE POSTULATES AND THE AXIOM OF UNIFORMITY.

CHAPTER X.

CERTITUDE AND TRUTH.

CHAPTER XI.

THE SEEN AND THE UNSEEN.

BOOK III.

ANALYSIS OF PHENOMENA.

CHAPTER VII.

ELEMENTS, ASPECTS, AND CONDITIONS.

§ 1. THE distinction of the two aspects in philosophy, and the analysis of the subjective aspect and its methods, have now reached a point at which we may profitably begin to make some application of the distinctions and principles therein discovered to philosophical reasoning upon phenomena at large. In the present Chapter accordingly I take for examination the distinction of elements and aspects, in contrast both with each other and with conditions, a term which is continually confused with them.

We have seen what is meant by the term 'object-matter of reflection,' and by what processes, and under what categories, reflection deals with this object-matter; namely, by conceptual processes, and under the categories classed as simple and modal. These are the methods employed by reflection. But it is not with these that we have now to do, at least in their character of categories and methods; it is with the application of some of them to the object-matter itself, and with their contradistinction from each other. To a great extent the present examination goes over old ground again; but with this new purpose in view, namely, to show what false routes are opened up by not adhering to the distinctions which we have already drawn, but confusing them with each other, or even forgetting them altogether. I begin then with the simple categories of element and aspect, which we have already made much use of, and proceed to distinguish them from another simple category, that of condition, as well as from each other.

I will first draw the distinctions anew, and define the terms which name them, then justify the definitions, and finally draw some of the consequences which flow from them. Premising, then, that we take phenomena as objects of reflection, and as percepts, that is, in their first intention, I state that we may take any such object in three ways; we may see, 1st, what its *constituents* are as perceived; 2nd, what character it bears *as a whole*, which no other object bears; and 3rd, with what other things, not itself, it stands in some definite time or space relations. It is evident that *anything* can be regarded in these three ways; we do not need to make the proviso that it must be an object of reflection in its first intention. We are applying a perfectly general method; but, in choosing this particular kind of objects to treat in this way, we are applying it to the special object-matter of philosophy. The constituents discovered in such objects are what I call *elements*; the peculiar and exclusive characters of the whole are what I call *aspects*; and the other things to which the one in question is related are what I call *conditions* of that object.

The whole of metaphysic rests ultimately and in the last resort upon this perceptual analysis of objects of reflection. The examination of elements and aspects gives the determination of the *Nature* of phenomena, and is a statical examination; that of conditions gives their *Genesis* or *History*, and is a dynamical examination. The questions in metaphysic are, to determine what are the *ultimate* elements of phenomena, what if any are the *universal* elements of phenomena, and what if any are the *ultimate and universal* aspects of phenomena. The determination of the *conditions* of phenomena is not strictly speaking a metaphysical question; and that for two reasons, the first of which is, that it cannot be made an universal question, but must always be a question as to the conditions of this or that particular object; for the sum of things, phenomena as a whole, can have no conditions outside itself, it is a contradiction in terms. The second reason, depending on this, is, that to take the question of condition *as if* universal, that is, to seek what conditionality or nexus itself consists in, is according to Ferrier's theorem :[1] *Frustra quæritur quomodo inter ea conjunctio efficitur, quæ non nisi conjuncta reperiuntur*, to seek an analysis of the ultimate constituents of objects, and therefore falls under the enquiry into *elements*.

If there is one thing more than another which must be fatal to any philosophical doctrine, it is confusion between these three well-marked categories of elements, aspects, and conditions, and the consequent reference of anything to one category when it belongs to another; elements or conditions are thus made to do duty as aspects ; aspects or conditions as elements;

---

[1] See Chapter VI. Vol. I. p. 436-7.

conditions as both at once; aspects or elements as conditions. We get empirical or ontological systems instead of metaphysical ones; systems which clump what ought to be analysed, and analyse what ought to be clumped. There are but two great divisions of professed philosophies, the empirical and the metaphysical. The ontological sort is a case of the empirical; only differing from it in having its entities more abstract and unpopular. The great Neo-platonist ontology had, in its τὸ Ἕν, a clumped object which ought to have been analysed. The empirical philosophy which takes Force as the Cause of Motion, or of Change in respect of Motion, has a clumped object, Force, which ought to be analysed. The same may be said of the philosophy which takes Matter, whether with or without motion, as the ultimate existence. Matter, Force, and τὸ ἕν, are each of them nothing but phenomena, isolated, and imagined as if existing absolutely in any case, ἄνευ τῆς ὕλης in some cases.

First, then, as to the ultimate and universal elements of phenomena. These are three in number, the element of Feeling, which we have called the material element; the element of Time or duration; and the element of Space, which primarily attaches only to the two kinds of feeling known as sight and touch, including or combined with sense of effort in muscle or nerve. The two latter elements, those of Time and Space, are called the formal elements.

The proof that these elements are ultimate and, subject to the above restriction in the case of space, universal elements in all phenomena, is in one sense coextensive with, and to be drawn from, all parts of experience. There is no experience which does not prove it. It is a case of not seeing the wood for the trees, if any one supposes himself not to see it. But the way to look for it, the guidance of the eye to see the wood in the trees, may be given as follows. Take any general phenomenon, no matter whether objective or subjective; say for instance an objective one, *motion;* whatever is a case of motion will be covered by its analysis, and must include the analysis of motion in its own. How is motion to be defined? or, in other words, *What is* motion? To define it you must take it subjectively, and ask what is perceived in perceiving motion. Subjectively, motion is change of feelings in respect of space. Those are its elements of analysis, which, put together in a term of extension, become its definition. But what is change? Different feelings replacing one another in time. Everywhere we get the same three ultimate elements, time, space, feeling. Change is not an ultimate *element;* it is an ultimate *concrete* or *empirical.* It is the ultimate empirical in time consciousness alone, as motion is the ultimate empirical in the form of space as well as time. And you cannot make change an ultimate *element* without violating Ferrier's theorem, for it *is* those two elements, time and feeling, taken together, and is nothing beyond these. They are its analysis and definition, not it theirs.

It would be waste of time in a work like the present, addressed in some sort *ad clerum,* to follow out in other instances the same proof. For everything may be brought under the categories of motion or of change, as cases of the one or of the other. Suppose, for instance, we were to define Music, Painting, Justice, Expediency, Law, Chemical Structure, Molecules, Atoms, &c. &c.;—all such common things, however abstract, are plainly analysable and definable as cases of change or of motion. The difficulty in philosophy has been in dealing, not with acknowledged existents such as these, but with the imaginary entities formed from them, their supposed *causes*, as for instance, Force, The Will, Freedom, Substance, Action. In dealing with such *entia phantasiæ* as these, the only method is that of reflection, namely, to ask *what we mean* by the terms; which is Aristotle's famous method, always to insist on being told, of this or that phrase, τί σημαίνει; and again in definitions, it is Aristotle's method to require that the τί ὄν shall be laid at the basis, as a genus, thus avoiding the fallacious " that which " in definitions. For a " that which " in definitions usually introduces an attempt to define the proposed *definiendum* by its effects or by its conditions, by what it *does* or *suffers* instead of by what it *is*.[1] I say *usually*, for the phrase cannot wholly be avoided; it is useful not only for brevity's sake where the genus is known, but also for introducing descriptions of *ultimates* which cannot be defined. But in definitions it should always be closely questioned, and Aristotle's rule applied. Unquestioned, there is no end to the *entia phantasiæ* which may be brought in by it, as supposed causes of phenomenal effects.

These three ultimate and universal elements of phenomena (subject always to the restriction as to space) are the *metaphysical* elements of phenomena. By this is meant that, being subjective, they are also inseparable though distinguishable; are not concrete and empirical separately, but only in conjunction with one another. They are the *further* analysis of the ultimate concretes or empiricals. There must be a

[1] Topica, Z. 5. p. 142, b. 22. δεύτερος, εἰ ἐν γένει κ.τ.λ.

conjunction of a formal with a material element as the analysis of every concrete or empirical object. Feeling is combined either with time alone, or with time and space together. Time and space, the two formal elements, make no empirical object without feeling. For the divisions or distinctions in both of them come only from feeling. Pure time and pure space have, so to speak, no hold of one another.

Time, space, and feeling, then, the ultimate metaphysical elements of phenomena, stand in the place not only of Aristotle's Categories, but also of what we may call his two ultimate categories beyond these, namely, Matter and Form, ὕλη and μορφή, or εἶδος. But with this difference, that matter is no longer an indefinite, merely potential, something, unknowable by itself, ἄγνωστος καθ' αὑτήν.[1] For it is now named by a word with a well-known *meaning*, Feeling. The two Aristotelian categories, matter and form, are not included in his famous Ten, but stand apart from them; these being so many modes of form, requiring an indefinite subject-matter upon which to be imposed, and that in just so many generically different modes as there are categories. The subject-matter is thus divided into ten different sorts, not indeed equally ultimate, some being reducible to others, but still generically different sorts of matter, or existents, ὄντα, each of which has its own laws of causation, and gives rise to a science devoted peculiarly to its investigation.[2]

The Categories, thus brought into connection with

[1] Metaph. z. 10. p. 1036, a. 9.
[2] Metaph. Δ. 6. p. 1016, b. 33 ; 7. p. 1017, a. 23 ; 28. p. 1024, b. 10-16 ; Γ. 2. p. 1003, b. 12 ; p. 1004, a. 5 ; E. 1. p. 1025, b. 7 ; Θ. 1. p. 1045, b. 27-33 ; I. 3, p. 1054, b. 27-31.

matter and form, are no longer a mere list of perceptual qualities or attributes, each of which is a general term or genus in logic, καθόλου ὡς κοινὰ, but are a list of genera or *classes* of empirical substantive existents, γένη τῶν ὄντων. Each category is the "form" of a peculiar "matter," and this matter is found in all the individuals belonging to the class, matter being conceived as belonging to the world of *intelligibles*, νοητὰ, as well as to the world of sensibles, αἰσθητὰ, and expressly included in *everything* except what may turn out to be a case of pure form, although substantial, namely, of the τί ἦν εἶναι.[1] The relation of genus to species in logic is analogous to this relation of matter to form in substantive existents, inasmuch as a common notion, that of the genus, is specialised into definiteness by the addition of differentiæ. But observe, (and this is just one of the most slippery places in philosophy), the parallel is not exact; for in substantive existents there are but two concurrents, matter and form, the form being prior to the matter as known to us: while in logic there are two concurrents, genus and differentia, and a third besides, the resultant species, of which those two together are the definition. In the latter case the union of the two concurrents is, in the former it is not, a third thing; there is a third thing in logic, but not in existence.

These considerations will throw some light on an ambiguity which has always been a puzzle to students of the Aristotelian Categories, an ambiguity which cannot be removed from them as they stand. It consists in their mixing up substantive existents with attributes or determinations, and consequently appearing now as a list of summa genera of existents, γένη τῶν ὄντων,

[1] Metaph. Z. 10. p. 1036, a. 8; 11. p. 1036, b. 35.

ELEMENTS, ASPECTS, AND CONDITIONS. 11

Book III.
Ch. VII.

§ 1.
Elements of
Phenomena.

and now as a list of summa genera of attributes, γένη
τῆς κατηγορίας, or καθόλου ὡς κοινά. And the ambiguity is irremovable because it is found in the name
of the first category, which is most frequently called
οὐσία, in which sense it is the special object-matter
of the Metaphysica; but is also very often called τί,
or τί ἐστιν, which involves an adjectival character.

Aristotle in fact used his list of Categories, (however he was led to them), for two different purposes,
which he was enabled to do because they were a perceptual analysis, or list of features in objects as percepts. As such, they stood related to Logic on one
side, to further analysis of percepts as existents on
the other side. The list of categories, in its relation
to logic, is opposed to the *logical* categories of genus,
species, proprium, accident; and in this use they are
a list consisting of existents (1st category) and nine
generic perceptual attributes (the other categories).
But as facing towards the further analysis of percepts
as existents, they are brought into connection with
metaphysic as a whole, and the special sciences which
hold under it, in virtue of the analysis of phenomena
generally into matter and form; and in this use they
are a list consisting of pure existents (1st category)
and nine genera of modified existents (the other categories), into which existence enters with special modifications, as *quantities, qualities,* and *relations.*

Thus there are, with Aristotle, *two* great orders,
the order of Existence and the order of Logic, with
different sets of principles, ἀρχαί, appropriate to each.[1]
The Categories are common to both. And the whole

---

[1] Metaph. B. 1. p. 996, a. 1. ὅτι αἱ ἀρχαί πότερον ἀριθμῷ ἢ
εἴδει ὡρισμέναι, καὶ αἱ ἐν τοῖς λόγοις καὶ αἱ ἐν τῷ ὑποκειμένῳ.
See also cap. 3. of the same Book, p. 998, a. 20 et seqq.

scope and purpose of the order of logic is to *discover* the order of existence. The τί ἦν εἶναι is the *logical* name for the ultimate οὐσία which lies at the root or heart of every substantive existent. And to discover the τί ἦν εἶναι in Metaphysic would be to discover the *pure* as well as ultimate οὐσία of existence at large, or *as existing*. There is thus a certain parallelism between the logical and existential analyses, which we may exhibit thus:

EXISTENTIAL.	LOGICAL.
οὐσία	τί ἦν εἶναι

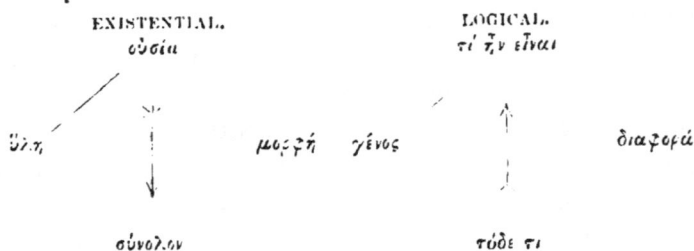

ὕλη	μορφή	γένος	διαφορά
σύνολον		τόδε τι	

It is worth while to remark, that while Aristotle had so far advanced beyond Plato, as to banish the χωριστὰ εἴδη from logic, he distinctly stopped short of banishing their parallel, οὐσία, and *a fortiori* pure οὐσία, from the order of existence. As remarked in a previous Chapter, he left causative agency *in*, and therefore predicable of, the πράγματα, while denying its separate existence παρὰ τὰ πράγματα, in the shape of εἴδη. This causative agency he conceived as residing in the οὐσίαι, and as derived ultimately from the οὐσία χωριστὴ, ἀκίνητος, ἀΐδιος, which was pure energy, *actus purus*, without anything merely potential, therefore without anything material, about it; the οὐσία of οὐσία itself, so to speak, was ἐνέργεια. And this was his solution of the great problem of metaphysic, or as he conceived it of Ontology.

Notwithstanding that his own analysis clearly led to the difference pointed out above between logical

ELEMENTS, ASPECTS, AND CONDITIONS. 13

Book III.
Ch. VII.

§ 1.
Elements of
Phenomena.

objects and substantive existents, namely, that only in the logical objects there is a *third thing*, differing from the union of the two elements; and notwithstanding that he had rejected this third thing in the case of logic, as a χωριστὸν εἶδος; he yet did not see that there was no third thing at all, but retained it as an explanation of substantive existence.

Now I maintain that my list of elements, Feeling, Time, and Space, fills the place and performs the functions of Aristotle's Categories; not only are they a list of ultimate, irreducible, perceptual attributes, but these attributes, in various combinations, give the ultimate kinds under which existents can be brought, as cases or instances under the logical categories of genus, differentia, and species. I mean that feeling and form are ultimately the two intersectors which stand as logical genus and differentia to each other, and so combined are a species. To some combination of them every perceived or imagined property can be reduced. They it is which, when fixed upon by volition, become concepts in predication and stand as genus and differentia to each other in terms of extension. This passing through the logical mill, so to speak, gives them back as *existents*, classifiable under γένη τῶν ὄντων, whereas they went in as determinations or attributes of existents, or rather (and more accurately) as features in percepts which afterwards, when we have got the notion of existent to compare them with, we call determinations of existents, classifiable under γένη τῶν κατηγοριῶν.

We are here upon ground far more abstract than the common psychological distinction between objects of sense and objects of reasoning. For we are beyond the commencement of judgment, and we traverse the distinctions between modes of feeling, such as the bodily senses, or these and the emotions, or between the various emotions and passions. In every moment of consciousness, however short or incomplex, there is, as shown in previous Chapters, a formal feature and a material feature distinguishable but inseparable, and the material feature is double, that is, consists of two different sub-feelings. And this fully bears out Aristotle's doctrine, that sense is a *critical*, that is, a *discriminating* faculty, though not a faculty of judgment.[1] Sense does not judge, but gives us the materials of judgment; but these must be incipient or inchoate judgments, or else they would not enable us to go on to complete ones. We *see* perceptual determinations which afterwards are either generalised and classified in thought, or else are referred by thought to previous classifications; the operation in the latter case being often so easy and habitual as to have long sunk below the threshold of consciousness. But this circumstance does not make it an operation of special and so-called intuitive sense; rather it is the reason for holding it not to be so.

In this light we must read the passage of the Posterior Analytics,[2] where Aristotle speaks of sense being "of the general." He is here describing a conceptual process, and points out at what step of that process sense comes in. Sense is neither of the universal nor of the particular, as such, because it is of objects as percepts and not as concepts, and these are

---

[1] De Anima, III. 2, p. 426, b. 8.

[2] Anal. Post. B. 19, p. 100, a. 15: πάντος γὰρ τῶν ἀδιαφόρων ἑνός, πρῶτον μὲν ἐν τῇ ψυχῇ καθόλου (καὶ γὰρ αἰσθάνεται μὲν τὸ καθ᾽ ἕκαστον, ἡ δ᾽ αἴσθησις τοῦ καθόλου ἐστιν, οἷον ἀνθρώπου, ἀλλ᾽ οὐ Καλλίου ἀνθρώπου) πάλιν ἐν τούτοις ἵσταται ἕως ἂν τὰ ἀμερῆ στῇ καὶ τὰ καθόλου, οἷον τοιοσδὶ ζῷον, ἕως ζῷον· καὶ ἐν τούτῳ ὡσαύτως.

conceptual distinctions. At the same time it is of both incidentally, κατὰ συμβεβηκὸς, because all its objects become afterwards universals or particulars indifferently. Although, says Aristotle, it is *Callias*, an individual, that is the object perceived, yet what sense perceives is not Callias, but general qualities, say the qualities that make up a human figure; we see certain modes of qualities, and putting them together *judge* that they are Callias. The words occur in that inestimable passage in which Aristotle endeavours to seize and depict the tentative process of induction, or inductive imagination, νόησις.

It is a point of the greatest importance that we have the elements of perception in combination with each other independently of any exercise of attention for the purpose of knowledge, and therefore prior to any exercise of thought or conception. Thought is not required to put scattered elements of perception together; the elements of perception are always and from the first in combination with each other. There are no such things as scattered elements of perception, unless it be that, in comparing any two ordered, non-scattered, series, one has less, the other more, regularity in its empirical parts. That the elements of percepts are not and cannot be wholly disorderly, scattered, or chaotic, is guaranteed by two of the three kinds of them being time and space, the formal elements. There must be *some* sequence, and *some* coexistence, in perception, were it only from the fact that feelings do not exist but in time, or in time and space together.

But it may be said by the "scattered sensation" theorist, that, granting so much, still the differents which come before us in perception, in sequence or in coexistence, may be given to us originally as chaotic sequences and coexistences, never repeated twice; and that thought is requisite to bring this chaos to order and regularity. This, I need not say, is the Kantian theory, the Kantian Categories of Thought being proposed as the instrument with which the Subject makes the phenomena orderly, and educes the phenomenal and natural world.

My reply to this theory, that the sequences and coexistences of differents in perception are chaotic unless and until they are reduced to regularity by a process of thought with conceptual machinery, is this: let us examine experience. What do we find? For my part I find this, that in presentative perception, say for instance in running my hand along this table edge, or say (in order to eliminate the personal action) in watching a fly run across a window pane, I *see* a series of changes of position all connected with one another, and leaving behind them in my memory traces which in representation form a curved continuous line across the continuous surface of the glass. All this *I see*, although in describing it I have to use language which implies that I have formed a notion of what continuity is, what surface is, what objects are, what existence is. It makes no difference to what *I see*, what my notion of continuity is, of surface is, and so on. What *I see* has an order and a regularity of its own quite independent of the classes of thoughts or things to which I afterwards refer it; and quite independent of the attention, for the purpose of knowing more, which I may bestow upon it afterwards in redintegration.

How and why it is reproducible in redintegration was shown in Chapter IV., namely, because every the least moment of perception, being a moment of change, contains a portion which is strictly speaking *past*, not present. And both its order in original presentation and its order as reproduced in representation are not products of, but a prior pabulum for, the exercise of volition in reasoning. The trains of spontaneous perceptual redintegration repeat the trains of presentative percepts. There is an order in the sequences and co-existences of differents in both, which precedes and is modified by attention in voluntary redintegration or reasoning, and, so modified, results in the conceptual order. The train of percepts is not disorderly in its elements, but partial and therefore disconnected as a picture of the world. And this partialness it is which we seek to remedy by the conceptual order.

There are three several respects in which percepts are orderly, independently of conception. First, in presentations of sight and touch, there is a shape or figure in which the sensations are presented. Secondly, in sequences where representation is involved with presentation, as in the case of the fly crossing a pane of glass, there is an order of the parts of the total image. Thirdly, the presentations and mixed presentations and representations are repeated in representations, and this repetition is itself an orderliness, precluding the notion of a sensation chaos in which nothing is recognisable as the same with what has gone before.

Physiology also lends a support to this view of the matter, by its discovery of the nerve machinery upon which the presentations and representations depend. And this is saying, in other words, that psychology furnishes a support to metaphysic, in comparing the genesis of states of consciousness with their analysis. Physiology, in fact, distinguishes the nerve action

which accompanies and supports sensations, centri-
petal stimulation of the organ, from the nerve action
set up centrifugally in consequence of this. The first
action is that which supports consciousness simply,
the latter that which supports volitional consciousness.
But no physiological action has ever been discovered
which in any way suggests that volitional action is
accompanied with *a priori* forms of thought; though
of course this absence of evidence for, is no conclusive argument against, the existence of such forms.
This is a question ultimately of subjective analysis.

If there were no such order in percepts and perceptual spontaneous redintegrations, what hold could
any pure concepts or categories of thought have over
the chaotic sensation content, which they are supposed
to form into objects and reduce to order? What
common bond could there be between them? It
would be necessary to go farther, and imagine the
sensation content to be not only ordered but altogether produced by the conceptual energy. And this
was in fact the post-Kantian ontologists' step forward
from Kant.

Again, if there were no order in percepts and perceptual spontaneous redintegrations, no *verification*
of the conceptual order would be possible, supposing
that order to have arisen in the *a priori* way contended
for by Kantians. There would be no "facts" to compare it with and test it by, but facts of its own creation, facts with no more testing power than the
thoughts which they were to test. Since however all
science is built upon verification, this theory according to which verification is an impossibility is a
theory incompatible with the existence of science in
any genuine sense. To obliterate percepts in dis-

ELEMENTS, ASPECTS, AND CONDITIONS. 19

BOOK III.
CH. VII.

§ 1.
Elements of
Phenomena.

tinction from concepts is to obliterate science in distinction from philosophy.

Casting a brief glance back at what has been urged with respect to the metaphysical elements of phenomena, we shall find that what we have done is this,—we have cleared these elements from the admixture of *conditions*, and that in two directions. First, we have got rid of the *condition* latent in Aristotle's ὕλη, and also in his τί ἦν εἶναι. We have done so by obtaining a list of ultimate elements in which complete empirical *existents* are not included; a list of ultimate genera of attributes. With him the τί ἦν εἶναι was at once the *cause* and the *nature* of itself. These two conceptions, cause and nature, which Aristotle identified, must be kept asunder; at any rate, they must be clearly understood as distinct in origin, and their analysis fully made before they are combined in philosophy. The nature of phenomena means their analysis; and the metaphysical analysis of phenomena is into their elements kept clear of conditions, or of the question *what makes* them be what they are.

Secondly, the elements have been cleared of conditions in respect of their putting together in thought. Whenever it is supposed that concrete percepts as existents require some special and separable energy of thought, some *a priori* concept or category, in order to be concrete percepts; that it is not enough to enumerate the elements of analysis, but that you must name the bond of combination between them, this bond being a special and separable energy of thought; then a condition is mixed up with elements, the *what makes* is brought in among them, and that in order to the *what is*. This clearance is effected by showing that the supposed bond of connection between elements is nothing else than one of the elements themselves,—namely, the formal element in one or other of its modes.

§ 2. To come in the second place to the consideration of Aspects. The term *aspect* has of late come to be used very frequently, and therefore as might be expected very loosely. It is necessary to define it with accuracy, with a view to its philosophical employment. As used by most writers it is a merely popular term, employed to signify that they are about to notice a different feature of the subject from what has gone before; it is used as equivalent to the term *side;* a new side of the subject in hand, or some new respect in which it is to be considered.

Aspect, as a philosophical term, means a character coextensive with and peculiar to the thing of which it is an aspect. It is in Aristotelian phrase an ἀντικατηγορούμενον, and ἀντικατηγορεῖται τοῦ πράγματος. Definitions and their *definita* are aspects of each other; the *definitum* is in order of intension precisely what the definition is in order of extension. But a single conceptual term is never the other aspect of a perceptual term: it is that perceptual term *plus* a modification for the purpose of reasoning. A definition is an instance of this purpose fulfilled; it is the full name of the thing defined by it, which has no other name in philosophy but its definition.

In philosophy there is one ultimate pair of aspects which are universal and necessary, Subject and Object; and this because reflection is the philosophical mode

ELEMENTS, ASPECTS, AND CONDITIONS. 21

BOOK III.
CH. VII.

§ 2.
Aspects of
Phenomena.

of consciousness. All other pairs of aspects are cases of this, and are tested as to their being truly aspects, ἀντικατηγορούμενα, by being exhibited as cases of reflection. This is what is meant by saying that all verification, in science as well as in philosophy, depends on reflection. To judge the equivalence, that is, the identity in difference, of two things, I must have passed back in thought from the second to the first, as well as forward from the first to the second.

The Postulates of Logic, which are a formulation of the principle of contradiction, are also a formulation of the aspect-making function of reflection. Reflection moves by the principle of contradiction, and formulates it in the Postulates. The postulate of identity, A is A, and the postulate of contradiction, No A is not-A, when so formulated by reflection become aspects of each other; and the third postulate, that of excluded middle, is an explanation of what is meant by the two former, namely, that each says the same thing in other terms, or that they are aspects of each other.

Having thus defined the philosophical sense of the term *aspects*, I come to the case of the two ultimate and necessary aspects in philosophy, the subjective and the objective. It is here that the greatest accuracy is requisite, and it is just here that the confusion of ordinary writers is most appalling. The high and abstract region in which this distinction arises is the watershed of philosophical systems. On one side, from the distinction rightly drawn, flow down the systems that may be true; on the other, from drawing it amiss, the systems that must be false.

Aspects must be kept clear of elements on the one

BOOK III.
CH. VII.
―――
§ 2.
Aspects of
Phenomena.

hand, and of conditions on the other. These are the two dangers to be avoided, the two subtil and insidious pollutions of the sacred well. Taking consciousness as consisting of the elements of time, space, feeling; existents as consisting of the same elements; and phenomena as consisting of the same elements; we have consciousness and existents as the two aspects, subjective and objective, of each other. And *phenomena* is a term for the union of these two aspects, being a word which expresses both, *things* that *appear, existents* in *consciousness*. Consciousness is the subjective aspect, existents the objective, and phenomena the two together,—a term expressing their inseparability. The case is an exact parallel of what was noticed above of the postulates of logic. Only that here we have in these expressions a full, concrete, perceptual content, whereas there we had unfilled logical formulas.

No single element constitutes consciousness, or existents, or phenomena. We must not make Feeling, for instance, into a representative of consciousness, alone and without the formal element; as would happen, if we regarded space as the objective element, or time and space as having a source separate from feeling. The elements are not to be separated from one another, and placed one in the subjective, the other or others in the objective, aspect. When Kant, for instance, assigns an *a priori* source to time and space, and an *a posteriori* source to sensible impressions, he divides the elements and distributes them into opposite aspects. Introducing the question of *source* is introducing the question of *conditions*. He has therefore to satisfy this order of enquiry, and go on at once to determine from which source objectivity comes. In Kant's *a priori* source, all Hegel is potentially contained. All separation of the ultimate elements of phenomena involves a confusion of the relative content of the aspects of phenomena; it therefore destroys them as aspects, and falsely exhibits them as conditions, of each other. The tendency of the mind to fall into this error consists, no doubt, in the readiness with which we ask after the cause or reason of anything so soon as we apprehend its denotation. Without waiting to be informed *what* is time, *what* is space, *what* is feeling, we rush to the enquiry *whence comes* our notion of time, of space, of feeling. The analysis of the familiar is neglected for the search after the unfamiliar. Curiosity is preferred to knowledge. The search after conditions, which is the life blood of science, is the poison of philosophy, which consists in analysis. And then the wolf complains of the lamb, and science accuses philosophy of polluting the stream of knowledge by throwing in the research into *causes*, a conception peculiar to science. The antidote to this tendency to introduce the question of conditions is found, not merely in pointing out its evil, but in what has now been done, namely, in showing how we may know when we are yielding to it. It manifests itself in a separation of the inseparable elements, and in a consequent confusion and destruction of the inseparable and coextensive aspects.

The mental confusion of aspects either with elements or with conditions is one which clothes itself in various shapes; for instance, in an inaccurate use of the term *factors* of cognition. A factor may mean either a constituent part of the factum or product, and then it is an element; or it may mean a pre-

requisite, an agent in its production, and then it is a
condition. The expression *factor* clumps the two
distinct notions of element and condition, and with-
draws them from attention and analysis.

Thus, for instance, we find Ferrier speaking of
object and subject now as factors, now as elements.
"Thus Knowing and Being are shown to be built up
out of the same elements." And "Plato's intelligible
world is our sensible world to this extent, that it is
that which *must* embrace a subjective and an objective factor—an ego and a non-ego."[1] From these
passages it would seem as if Ferrier recognised no
distinction between elements, aspects, and conditions.
Factor seems to cover all alike.

Mr. Lewes, in one passage, uses the term factor
apparently with the intention of identifying elements
and conditions. "Experience," he says, "is the
registration of feelings and the relations of their
correlative objects. Science is the explanation of
these feelings, the analysis of these objects into their
components and constituents, which are then held to
be the factors of the facts. These factors are of
various kinds, real and ideal, concrete and abstract,
appreciable by Sense, and appreciable by Intuition."[2]
But in other passages *factor* need not mean more than
condition only. Thus Mr. Lewes speaks of light as
" the product of the undulations of the ether, and of
retinal sensibility. Both factors are indispensable to
the product, &c."[3] At another place he speaks of
" cosmical factors as coöperant with sentient factors
in the production of these feelings," those of sound.[4]

---

[1] Institutes of Metaphysic. Ontology, Prop. X. Obs. 4 and 8.
[2] Problems of Life and Mind, Vol. I. p. 100.
[3] Same Work, Vol. II. p. 235. [4] Same Work, Vol. II. p. 173.

ELEMENTS, ASPECTS, AND CONDITIONS. 25

Book III.
Ch. VII.

§ 2.
Aspects of
Phenomena.

Very true; they are factors in the conditions, but not elements in the products, light and sound. Still it would have been better, if this had been distinctly remarked.

The reason for its not being distinctly remarked lies probably in an inadequate analysis of feeling. Mr. Lewes recognises no distinct elements in the minima of feeling, but considers each feeling as an unit, with a certain "signature;" the time and space elements being ignored as elements and clumped with the units of feeling, as their signature or quality. It is a subjective repetition of the *materia signata* of St. Thomas Aquinas.[1] "Knowledge begins with indefinite Feeling, which is gradually rendered more and more definite as the chaos is condensed into objects, effected through a rudimentary analysis determined by the fundamental Signatures (Qualities) of Feeling, namely, Tension, Intension, Extension, Duration, Likeness, Unlikeness."[2] And again, speaking of extension, "This is undoubtedly a property of Matter which, because it is one of the fundamental Signatures of Feeling, cannot be thought absent."[3]

Let it not be imagined that this is merely a verbal difference, or that it is over refinement to insist upon the formal elements of states of consciousness being called elements and not signatures of feeling. The term element expresses the fact that empirical feelings can be subjectively analysed; whereas to take them as feelings with signatures is to take them as unanalysable units. And then the consequence follows, that psychology and physical science can give a fur-

[1] See above. Chap. V. Vol. I. p. 313.
[2] Problems of Life and Mind. Vol. I. p. 101.
[3] Same Work, Vol. II. p. 278.

ther analysis, namely, an analysis of the conditions of what are to metaphysic unanalysable ultimates. Whereas the truth, which is expressed by using the term elements, is, that metaphysic can analyse subjectively the ultimates of psychology and physical science, even when those ultimates are proposed as conditions of feeling.

The distinction between aspect and condition gives us the answer to an objection which is often proposed as a *crux* of the metaphysical doctrine, that *esse = percipi*. On this hypothesis (it is asked) what becomes of objects when no one is present to perceive them; of the lamp, for instance, when I leave the room and no other person is there; must not the lamp, on your hypothesis, cease to exist? The answer is, that the objection confuses aspects with conditions, by taking a percipient as a *conditio existendi* of the lamp. The lamp exists, even when no one is presentatively perceiving it, if its conditions exist or have existed. These conditions are, many of them, representations; and so is the lamp itself, when I have left the room. What is meant by the *esse* of the lamp being its *percipi* is, that, except for perception, presentative or representative, I should have no knowledge whatever of the lamp, neither of its *whatness*, nor (*a fortiori*) of the fact *that* it exists. It would be for me non-existent. But the *esse* of the lamp is satisfied by the mere perception of what it is, that is, of what the term means to my mind.

The confusion between aspects, on one side, and either elements or conditions, on the other, underlies and pervades the whole of modern German speculation, Leibnizian, Kantian, and post-Kantian. In the Leibnizian Monads we have it in its most palpable

ELEMENTS, ASPECTS, AND CONDITIONS. 27

Book III.
Ch. VII.

§ 2.
Aspects of
Phenomena.

shape. The Monad is a simple substance exercising perceptive and appetitive powers,[1] a conscious and active existent. The "simple substance" represents the objective aspect, and becomes the condition of the perceptions and appetitions which represent the subjective aspect. The two aspects are rolled into one by making one of them the condition of the other.

Sometimes this same admixture is adopted, but with a modification which certainly renders it more logically consistent, since the objective aspect is made the starting point. The "substance" is endowed not with a perceptive but with a self-manifesting power. Thus Herr Fechner says: "A mind comes to light and comprehends itself without intermediate aid." (*Ein Geist erscheint und erfasst sich unmittelbar selbst*).[2] And in another passage, where this view is farther enlarged upon, we read: "The same Being (*Wesen*) has two sides, a mental, psychical, side so far as it is capable of coming to light (*erscheinen*) to itself, and a material, bodily, side so far as it is capable of coming to light to others, in another form; but body and mind are not bound together as if they were two essentially different Beings."[3]

Striking as is the turn thus given to the matter, especially in the explanation of what Body is, and what Mind is,—namely, manifestation *to others* and manifestation *to self*,—I do not see that it escapes the Leibnizian fallacy. For, unless we imagine an existent, an objective aspect as agent, exercising the power of self-manifestation, we have no more in the concep-

[1] Monadologie, §§ 1 to 20.
[2] Zendavesta, Vol. I. p. 410.
[3] Same Work, Vol. II. p. 322.

tion of self-manifestation than we have in that of self-perception, or reflection. Now it is just this existent Being that Fechner's words bring before us; and the union of aspects is saved on this theory by the same expedient as in Leibniz, by making one the condition of the other; only that the nexus between them is conceived as power of manifestation, an *Erscheinungs-kraft*, and not as power of perception.

The same or a very similar conception meets us in Dr. MacVicar's Sketch of a Philosophy. In his Introduction to the First Part of this work, Dr. Mac-Vicar, after having in two Chapters acutely and forcibly contended for the method of reflection in philosophy, and shown the inadequacy of the "so-called scientific method, or method of merely outside observation," comes in the third Chapter to enquire into the relation of consciousness to pure intelligence, and in the course of that enquiry, to ask what are the essential traits of Being, taking the term in the largest possible sense. His answer runs partly in these words:[1] "May we not say that all Being is essentially rela-tional in this respect, at least, that it is self-manifest-ing to other Being—not, indeed, as outwardly per-ceptible object to all other Being—not as outwardly perceptible object to such defective recipients of self-manifesting power as we ourselves are—but self-manifesting, inwardly or outwardly, or in some way or other, and more or less, to all other Beings and things, and perfectly self-manifesting to a Being who possesses perfect perceptivity, such as God. Yes; self-manifesting power is an essential attribute of everything that exists. This is proved by the very

---

[1] Sketch of a Philosophy, Part I. p. 30. Williams and Nor-gate, 1868. The work has been since completed in Four Parts.

conditions under which alone existence can be admitted by us. In forming a notion of any Reality, however much we may strip it of all its other properties, we must leave it in possession of possible perceptibility, or a self-manifesting power. The moment we deprive reality of possible perceptibility it can be held as a Reality no longer."

Here we have as it seems to me the confusion between aspects and elements in the most palpable form. "Possible perceptibility or a self-manifesting power." For perceptibility is an *aspect*, but self-manifesting power, as an "attribute," is an *element* in the analysis of Being, and the *condition* of its being actually perceived. We may hold that *esse* = *percipi*, as a matter of "nature" and definition, without in the least being called upon to admit that percepts are endowed with a self-manifesting power as their essential attribute. The phrase self-manifesting power is an attempt to carry us back, behind its definition, to the condition of existence at large, by substituting an element or attribute of existents for the existents themselves as objective aspects. And this would inevitably lead the author himself back to that very world of separate, empirical existents, and so-called scientific methods, out of which we have seen him striving to escape into the world of philosophy.

For be it noted that " self-manifesting power," as the term is used by Dr. MacVicar, is no mere amplification of self-manifestation as a simple fact. It means much more than this. It intends to insist on a *power* or *force* which issues in self-manifestation. This is clear from a later passage, where he says, speaking of the term *existence:* " It is the standing-out-ness or self-manifesting power of the object, not merely as anything in general, or as anything different from what it is, but as the very thing that it is. We do not go the length of those who maintain that the whole nature of that which exists is to be perceived, so that an object cannot exist but when it is perceived, its existence being constituted by its being perceived. Existence we regard not as 'idea' merely, but as 'force.' And that force we regard as essentially self-manifesting, or spontaneously radiant, so to speak, into that which is 'idea,' if there be a recipient of ideas, or a percipient of ideas, or more generally a percipient within the sphere of its action. This view, as it appears to me, is fully borne out by the information of thought itself as to the nature of existence. I find that I can think away all the other attributes of substance; but immediately I attempt to suppose the existence of that which cannot manifest itself in any way, that which is not perceivable by any percipient however perfect, then to me that vanishes altogether—it ceases to be."[1]

We find this same fundamental conception again in a work of the greatest ability and deserved celebrity in Germany, Herr Professor Lotze's *Microcosmus*. "Only as forms or states of intuition or cognition can the content of sensibility, light and colour, tone and odour, be comprehended; but if these are to be not only phenomena in our mind but also inherent attributes (*eigen*) of the things from which they appear to issue, then the things must be capable of coming to light to themselves (*sich selbst erscheinen*), and producing these attributes in themselves, in their own

---

[1] Sketch of a Philosophy, Part IV. p. 54-55. I shall have occasion to recur to this deeply interesting and able book in another connection. See Chap. XI.

sensibility. To this consequence, which spreads the clear light of living animateness over all Existents (*alles Seiende*), our longing must resolutely proceed; this alone can enable it to give actuality to the sensible without us, by giving it an actuality in the inmost nature of external things (*im Innern der Dinge*). Vainly should we attempt to attach to nonsentient things as an external property that which is only thinkable as an inner state of some process of feeling (*Empfindens*)."[1]

Lotze's reasoning appears to be—If the physical world is *real*, it must be endowed with sensibility, or power of manifestation to itself, since that is the mode in which the spiritual world is real, and the only mode of reality known to us. The conception of a self-manifesting power is the same as in the other passages quoted. If however my remarks hold good, this argument breaks down; of course without disproving the doctrine that all physical matter is endowed with consciousness. That doctrine may be true, or it may not; but it cannot be made to follow from any strictly metaphysical premisses; it is not a doctrine of metaphysic. Wherever it appears as a doctrine of philosophy, it is either obtained by a confusion between aspects and conditions, as in the cases which have been examined, or it belongs, as an hypothesis, only to the constructive branch of philosophy, not to metaphysic, the universal and analytic branch of it, in which it is an unwarranted intrusion.

When Idealist schools of philosophy fall into this confusion, it is usually in a more subtil way than by assuming a substance to begin with; they begin by assuming an action, the action of Reflection. But

[1] *Microcosmus*, Vol. I. p. 385.

what is the gain of changing the *venue* to Reflection, and taking it as an action, if you immediately objectivate the *subject* in that action? This is just what what Fichte does. It is shown by his use of the word *setzen, positing.* For instance, near the beginning of the Third Part, the Practical Part, of his chief work we read:[1] "The Ego posits itself absolutely and without any further ground, and it *must* posit itself, if it is to posit anything else: for that which does not *exist* cannot posit anything; but the Ego exists (for the Ego) absolutely and solely through its own positing of itself." The *Ego*, then, *is*, is an existent, before it can "posit" anything at all. The agent is the condition of its action, and that action is consciousness. Again farther on: "But the Ego, just because it is an Ego, has also a causality upon itself; the causality of positing itself, or the capability of Reflection (*die Reflexionsfähigkeit*)."[2] The subjective aspect, the *Ego*, becoming its own condition (the *causa sui* fallacy in another dress), clumped with causality, explains everything else. The only way in which such a highly abstract personage can move is by means of the most abstract of all distinctions, the logical postulates; and accordingly this is the way in which Fichte's *Ego* moves, as we see by the opening pages of the work which I have quoted. Hegelianism lay here already quarried to Hegel's hand; he had only to take the *Ego* and its method of movement by contradiction, and generalise it, make it a world-Ego, instead of a particular Ego; which accordingly he did.

[1] *Grundlage der gesammten Wissenschaftslehre.* Werke, Vol. I. p. 251.

[2] Same Work, Vol. I. p. 293.

But perhaps the writings most fruitful of insight into the mechanism of the fallacies which cluster round the distinction of aspect and condition are those of Schelling. Schelling was a man who laid down in a series of writings a record of the several stages through which his mind successively passed in perfecting his idealistic system. One of the clearest and completest of these records is the *System der gesammten Philosophie und der Naturphilosophie insbesondere*,[1] belonging to the year 1804, and published from his manuscript remains. He begins that work with a statement of "the one sole Pre-supposition which is forced upon us only by Reflection upon knowing itself." It is this: "The first pre-supposition (*Voraussetzung*) of all knowledge (*Wissens*) is— that it is one and the same thing that knows and that is known." He proceeds first to explain, and then to prove, this proposition. In explaining it, he says: "In the first Reflection on knowing itself we believe that we distinguish a Subject of the knowing, or the knowing itself taken as act, and the Object of the knowing, the thing known. I say advisedly *we believe* that we distinguish, for the reality of this distinction is the very thing in question at present," &c. He then proceeds to take this distinction, the reality of which he denies, as if it involved and imported a *separation* of the two aspects, and not merely a distinction between them as inseparables; though this latter import would seem to be well described by the words he himself uses, "*the knowing itself taken as an act, and the object known.*" He *neglects* this mode of distinguishing the aspects, and proceeds to argue against *separating* them just as I have argued against

[1] Werke, Abtheilung I. Vol. VI. p. 131.

it in Chapter II. He argues that to *separate* the aspects is to make them into *conditions* one of the other; and that therefore they cannot be separated.

Neglecting then, I repeat, the view of a distinctness between inseparable aspects, which nevertheless he gives as in accordance with the natural "belief" of reflection, he proceeds to adopt what he takes to be the only alternative of the separatist view; namely, that the two aspects form together one and the same thing, *dasselbe Eine*, whereby the two aspects cease to be aspects of each other, and become elements in a single existent, an existent which, notwithstanding that it is the *All* as well as the *One*, yet behaves as ordinary existents do, existents which are objects of *direct* perception, only on a larger scale; that is, it acts, perceives, and reasons, by means of the postulates of logic, and produces the objects of the natural world. Schelling's philosophy, in the form it assumed in 1804, was an Ontology based on taking the two aspects as if they were elements of an existent, which existent was the Universe.

It is worth while to notice the particular train of thought by which this sole existent is set in operation, and what its motive force consists in. Consider the following passage: "There is but *one* knowing, and *one* known. It follows from this that (§ 3.) The highest cognition is necessarily that, wherein this Identity (*Gleichheit*) of Subject and Object is itself cognised; or, (since this Identity consists just in this, that it is one and the same thing that cognises and is cognised), that, wherein this eternal Identity cognises *itself*." This reasoning, says Schelling, is self-evident and needs no proof.[1]

---

[1] Werke, Abtheilung I. Vol. VI. p. 141.

ELEMENTS, ASPECTS, AND CONDITIONS. 35

Book III.
Ch. VII.

§ 2.
Aspects of
Phenomena.

It certainly is so, *on the assumption* of one and the same existent, as knowing and known. But this assumption is as much precluded by reflection, to which Schelling appeals, as the assumption of a separation. Reflection does *two* things, distinguishes *and* connects. To insist on distinction alone is to make it separation; to insist on connection alone is to make it identity. Schelling does the latter. But in either case the *reflectiveness* of the proceeding is abandoned; and, while professing to appeal to reflection as the highest mode of consciousness, you are really falling back into direct perception. In this way, whether it is as One Thing, or as the Identity in that one thing, or as the Two Identical Terms of identity, we get a *conscious existent*, an object of direct perception, just as in the case of Leibniz, Fichte, Fechner, MacVicar, and Lotze; only in this case the method is more manifest, namely, by using aspects as elements, while distinctly (and rightly) rejecting them as conditions.

For be it observed, (and this is the root of the matter), it is not sufficient to constitute an Existent, that two opposite *aspects* are put together. For that you need *elements*. Otherwise there is nothing of which the aspects are predicable, or of which they are aspects. The requisite "something," the existent, which has the double aspect, must first be constituted; and this it is by its constituent elements. The first things which are used to define an existent become necessarily considered as its elements. And if the double aspects are used for that purpose, they *eo ipso* are denaturalised as aspects, and figure as elements. Their original relation to each other is destroyed, and replaced by a new relation; and the whole, the existent, which they constitute, sinks back again into the position of being one aspect of the pair, becoming an *object* to reflection; and inasmuch as the subjectivating power of reflection has been already employed in constituting the object, this object has the false character of an *absolute*, that is, of an existent as envisaged by direct and not by reflective perception.

The same, too, is the main error in Ferrier, the confusion of aspects with conditions under the notion of elements. His confusion of elements and conditions under the common term *factors* was noticed above. This seems now to depend upon a further confusion between aspects and elements. For want of this distinction he, too, comes to an Absolute Existence, which he thus defines : " Absolute Existence is the synthesis of the subject and object—the union of the universal and the particular—the concretion of the ego and non-ego ; in other words, the only true, and real, and independent Existences are minds-together-with-that-which-they-apprehend."[1]

Here again perhaps it will be said—Why object to this language ? Is not the difference infinitesimal between an Absolute constituted by two inseparable elements and an Existence knowable under two inseparable aspects ? Not so, I would reply. Apart from the difference of the analysis which leads to the one and that which leads to the other, a difference which has other consequences than the present, the conception of an existent composed of elements is smaller than the conception of one composed of aspects. For an existent composed of elements is but *one half* of the whole object of reflection. In the one case reflec-

---

[1] Institutes of Metaphysic. Ontology, Prop. X.

tion names one aspect of its whole procedure, in the other it names the procedure as a whole. Applying this to the Universe, we get in the one case, i.e., if we represent it as an Absolute composed of Object and Subject as *elements*, a limited horizon, an object not infinite propounded as infinite. In the other case, that is to say, if we take the Universe as existence knowable under two aspects, each of which consists of elements, infinity itself is before us. It is no longer treated as a rounded-off whole. It is one thing to say that a tower is roofed only by the sky; another, that the sky is the roof of the tower. The one lifts the tower to the sky, the other brings down the sky to the tower.

Ontological philosophies, therefore, which exhaust reflection in *constituting* their object, are essentially the same as empirical philosophies, or scientific theories proposed as philosophical; for both alike place an object of direct perception in the position which ought to be occupied by an object of reflective perception, an object with the two inseparable aspects, subjective and objective. The only difference is, that while the scientific pseudo-philosophies do not trouble themselves about the phenomena of reflection at all, the ontological philosophies professs to transcend reflection and arrive at a cognition of the Absolute. But in both alike the object is an absolute, in the sense of being an object of direct and not of reflective perception, an object self-existent in the sense of existing whether or not primary or direct consciousness of it exists also. Reflective consciousness, taken alone, becomes an absolute existent.

Not only, then, does reflection determine that there shall be philosophy, but the nature of reflection determines its nature, just as the nature of consciousness determines that of science. The confusion of aspects with elements arises from overlooking a fact in the nature of reflection, namely, as shown in Chapter II., that reflection is not a primary but a derivative mode of consciousness; arises from an ignorance which analysis of the phenomena of reflection would have removed. We have just seen how Schelling disregards that analysis, and leaps to the conclusion of the aspects forming a single existent. This he could not have done, if he had seen that reflection is not a direct perception of self, arising without antecedent experience of feelings in combination with one another; but an enrichment of a previously existing train of such experience by a new feature, the double character, objective and subjective, borne by those feelings. The elements of phenomena are given in the antecedent experiences; their aspects arise in reflection upon them.

Farther it must be recalled, that, unless and until the act of reflection arises, nothing can possibly be known *as existing*. For only in and by that act do we perceive that *esse = percipi*, that existence and perceivability are convertible terms, ἀντικατηγορούμενα. Note carefully, too, that the act of reflection does not tell us that *esse = percipere*, but that it = *percipi*. *Percipere* expresses any act of perception, either direct or reflective. If you express the act of direct perception in a proposition, say A is B, then the act of reflective perception, following on this, gives you two equivalent determinations of the *predicate* of the direct proposition; *i.e.*, B may be taken either as *esse* or as *percipi*. This is to take reflection as derivative and not primary. But if you take reflection as primary, and A is B as a proposition expressing an act of reflective perception, then A, the subject of the proposition, stands for the percipient Subject, and the percipient Subject is identified both with B as *esse* and with B as *percipi*. In other words, we have, as three convertible propositions, *esse = percipi; percipere = esse; percipere = percipi.*

In the former mode, reflection being supposed derivative, we have Existent = Percept; in the latter, reflection being supposed primary, we have Percipient = Percept. But the analysis of the phenomena of reflection by reflection itself is the final and decisive criterion in philosophy; and this analysis, by showing that reflection is derivative not primary, shows in the present case the falseness of the theory that *esse = percipere*, and consequently the falseness of ontological systems which are built on that foundation. Reflection, taken as primary, includes simple perception in itself; taken as derivative, excludes but supposes it. It is all-important to analyse reflection, and not to clump simple perception with it under one name. For in so clumping the two, you get a highly complex act, which nevertheless you lay down as the ultimate and unanalysable basis of philosophy, under the name of Reflection.

Now it is impossible to take *esse = percipere* without doing this; for to take *esse = percipere* is to assume an existent in the *esse*, and to turn *percipere*, an act, into *percipiens*, an agent. And this, again, cannot be done without identifying the subject of propositions, a term of pure logic, with the conscious Subject, a term of philosophy or applied logic. All this, however, must be done, if done at all, in the teeth of the analysis of reflection, as known to us by actual experience.

Observe, finally, the close intertwining of the two errors with each other. Take aspects for *elements*, and the result is to turn them into conditions; the subjective and objective aspects, clumped to form an existent, give you an agent clumped with its act, a conscious agent or a percipient; which is a form of what has been called Monism. Take, on the other hand, aspects for *conditions*, and the result is to give you two such existents, one a source of consciousness, the other a source of existence; which is a form of what we may call Dyadism. Take, however, the two aspectual conditions, formed as just stated, and consider them as reciprocal conditions of each other, and the result would again be a kind of Monism, in which the error committed by separating aspects was repaired by the opposite error of identifying conditions.

§ 3. We are thus launched upon our final enquiry, the consideration of conditions, the third member of our distinction. It would seem at first sight as if there was hardly a niche in philosophy where a new heresy could obtain a foothold, so numerous have been the heresies already propounded and accepted by different sections of the philosophical world. Nevertheless Mr. Lewes appears as the propounder of a new one, and that with great ability and wealth of illustration. His Chapter on Motion as a Mode of Feeling[1] may be regarded as an argument in support of that new doctrine of Monism briefly described at the conclusion of my last paragraph. This is the last Chapter of the book, as well as of his Problem VI.,

---

[1] Problems of Life and Mind, Prob. VI. Vol. II. pp. 456 to 504.

which is entitled "The Absolute in the Correlations of Feeling and Motion."

His argument assumes as one of its bases "the truth of the doctrine enunciated in Problem V., namely, that the logical distinction between the conditions of a phenomenon and the phenomenon itself is simply an artifice, there being not two things, a group of conditions (causes) on the one side, and a result (effect) on the other, but one thing differently viewed. What we call the conditions are just the analytical factors we have detected in the fact."[1]

Basing himself on this, Mr. Lewes goes on to prove that Motion is a mode of Feeling.[2] If this was all, I at least should have no objection to make. It is what I have always maintained.[3] But Mr. Lewes maintains also something else quite different from this, namely, that in the phenomena of sensation "the neural process and the feeling are one and the same process viewed under different aspects. Viewed from the physical or objective side, it is a neural process; viewed from the psychological or subjective side, it is a sentient process."[4] In other words, he is going to show that a neural process is not only the condition but the objective aspect of one and the same train of feelings. He is going to subsume a case under his general law of Problem V., at the same time that he shows this case to be one of inseparable aspects.

What I shall do then is this,—I shall show that conditions and their results are one thing, obverse

---

[1] Problems, &c. Vol. II. p. 460.   [2] Same place, p. 457.
[3] Time and Space, p. 83 et seqq.   The Theory of Practice.
Vol. I. § 6.
[4] Problems, &c. Vol. II. p. 459.

aspects another; and I shall take the facts as I find them stated by Mr. Lewes. "There is absolutely no evidence that this neural process precedes and produces its sensation."[1] But it is not requisite to a condition that it should *precede* its effect; it is enough that its effect is *dependent* upon it; which relation we express by placing it *mentally* before its effect, under the name of an invariable antecedent. Provided there is dependence, the condition may be either antecedent or simultaneous with its effect in order of existence.

For instance, a table supports a book; the table is a condition of the book being held at a certain height from the ground; lower the table, and the book is lowered with it. The table is *thought* as a *prior* condition, but exists as a simultaneous condition of the various heights of the book from the ground. The height of the book is not the same thing as the table. So also in Aristotle's often used instance of an eclipse of the moon. The passing of the earth between the sun and the moon, so as to intercept the sun's rays, is the condition of the eclipse, the darkening of the moon's disc, and is simultaneous with its effect. On the other hand, the making of bricks is a condition precedent to the building a house with them. No one will contend that these processes are one process differently viewed, or that they have the same analysis. They are plainly former and latter portions of one and the same longer process. Conditions, then, may be either precedent or simultaneous; the dependence of the conditioned on them is what is essential. Accordingly the neural process may be simultaneous with the sentient process with-

[1] Problems, &c. Vol. II. p. 466.

ELEMENTS, ASPECTS, AND CONDITIONS. 43

BOOK III.
CH. VII.

§ 3.
Conditions of
Phenomena.

out ceasing to be its condition, that is, thought as its invariable antecedent. Nor is our so thinking it a mere "artifice," but represents the necessary *depend-ence* of some phenomena on others, when they are broken up into particular and separate objects. It is forced upon us by that separation.

"So with the supposed transformation of a neural process into a sensation. The process is the objective aspect of the sensation. Instead of our feeling the sensation of sound, of colour, or of fragrance, we are mentally looking at the changes in the sensory organ."[1] Mr. Lewes means, I suppose, that, in mentally looking at the changes in the sensory organ, we *forget* that these changes were *felt* as sounds, colours, or fragrance. However, here is the point on which I venture to differ. Mr. Lewes has *two* subjective aspects for the same neural process. The changes in the sensory organ are *motion*, a motion of solids; motion of solids is an object which is analysable into certain modes of feeling, Mr. Lewes' "optico-muscular sensations," my "feelings of combined sight and touch changing in space." This subjective analysis of the motion of solids is the subjective aspect of the changes in the sensory organ. This is *one* subjective aspect. Mr. Lewes has another besides. The feelings *of any kind* attached to those changes, which arise in us when they take place; feelings which may be of *any* kind, sounds, colours, fragrance, emotions of various sorts, &c.; and are not restricted to the particular kind of "optico-muscular." Mr. Lewes clumps these two sets of feelings together under the single name of the subjective aspect. He has two subjective aspects for the same neural process; one being

---

[1] Problems, &c. Vol. II. p. 468.

Book III.
Ch. VII.
───
§ 3.
Conditions of
Phenomena.

the feelings into which it is analysable, the other those which arise concomitantly (or *possibly* subsequently) with it; one its *analysis* as motion, the other its *conditioned* feeling: one, itself subjectively, the other, some feeling of fragrance. sound, and so on.

This latter feeling, or set of feelings, it is, which is dependent upon the motions in the neural process. The dependence is a fact proved by experience. If you want to have the sensation of fragrance, for instance, you must set up some mode of motion in communication with the nerve of smell. And that mode of motion in the stimulus and in the nerve together is said to produce the sensation of fragrance; by which is meant, that the sensation of fragrance depends upon the motion.

When, therefore, Mr. Lewes objects,[1] that "to say that it is a molecular movement which produces a sensation of sound is equivalent to saying that a sensation of sight produces a sensation of hearing," namely, because the molecular movements which condition sensations are movements which we could see and touch. if properly placed;—there is nothing to be scandalised at, except the paradoxical shape into which Mr. Lewes is pleased to throw the proposition. It is merely saying, in other terms, that motion of solids can be analysed into modes of feeling. But it is only *in combination* that these modes of feeling are motion of solids, and therefore only in combination that they are *conditions* of other feelings.

Mr. Lewes, however, on the same showing, must admit a view not merely paradoxical in point of form, but logically suicidal; he must admit that to say that a molecular movement is the objective aspect of a

───
[1] Problems. &c. Vol II. p. 481.

sensation of sound is equivalent to saying that a sensation of sight is the objective aspect of a sensation of
hearing. He must admit one kind of sensation as
the objective aspect of a different kind of sensation,
and both alike as subjective aspects of one and the
same movement.

At one place Mr. Lewes seems on the point of
becoming aware of the two subjective aspects which
he assumes.[1] "The phenomenon known objectively
as a nervous tremor, a neural process involving very
complex elements of molecular energy, does not *become* a feeling in the sentient organism, it *is* that
feeling in the organism, and is the occasion of a quite
different feeling in the observer." Here, the feeling
conditioned by the neural process is identified with
that process, its condition; and the true subjective
aspect of the neural process is put aside, by calling
the process "the occasion of a quite different feeling
in the observer."

The "observer" occupies the place of reflection;
and it is reflection which distinguishes the objective
and subjective aspects. The modes of feeling into
which reflection analyses motion are the subjective
aspect of motion. And these are quite a different
thing from the modes of feeling which accompany
motion in the neural process. From the point of
view of reflection, or Mr. Lewes' observer, both these
latter things are seen *together*, accompanying each
other, and one depending on the other, the modes of
feeling on the neural process; they are not seen as
each taking the other's place, as its opposite aspect;
besides which, the neural process has, for the observer, a subjective aspect as a mode of feeling. The .

[1] Problems, &c. Vol. II. p. 488.

two sets of modes of feeling, thus attaching to the neural process, cannot be clumped as one set, to be called now effect, now subjective aspect; but these two names must be used to keep them separate, by being separately appropriated.

Mr. Lewes, then, has failed to show from the facts which he adduces, that the subjective aspect is identical with the objective. And he fails, because he confuses aspects with conditions, notwithstanding his law of identity of cause and effect. For, taking what he calls the subjective aspect, the feeling which he says *is* the neural process, (and which I call the feeling conditioned by it), he has separated this feeling from that which he calls " the different feeling in the observer," (which is the true subjective aspect of the neural process). He has provided himself unwittingly with *two* subjective aspects of one neural process. And the result accordingly is the very reverse of what was intended. For Mr. Lewes has not only not shown that a particular case of a condition, motion, is identical with what it conditions, *viz.:* the feelings which accompany it, but in trying to show it, by introducing the identity of the subjective and objective aspects, he has sundered the aspects themselves. Such is the consequence of identifying the objective aspect with a condition, by making it the condition of the subjective. Had Mr. Lewes seen clearly that aspects are inseparable and coextensive solely because they are products of reflection, he need not have mixed up his principle of identity of cause and effect with his attempt to prove that the particular case of motion and mode of feeling was a case of inseparability of aspects.

I have examined this doctrine as it appears in Mr.

ELEMENTS, ASPECTS, AND CONDITIONS.   47

BOOK III.
CH. VII.

§ 3.
Conditions of
Phenomena.

Lewes' Problem VI., Chap. IV., because it is this Chapter that contains all that even looks like a proof of it. But it is not the only place in which the identity of neural process and concomitant sensation is maintained. We have already, under the head of Psychological Principles, been told of the " Psychoplasm," from which the psychical organism is evolved, as the vital organism is from the " Bioplasm."[1] The " neural tremors" of the Psychoplasm are " the raw material of Consciousness." " The movements of the Psychoplasm constitute Sensibility." Of the mental organism we are told, " Here also every phenomenon is the product of two factors external and internal, impersonal and personal, objective and subjective. Viewing the internal factor solely in the light of Feeling, we may say that the *sentient material* out of which all the forms of Consciousness are evolved is the Psychoplasm incessantly fluctuating, incessantly renewed. Viewing this on the physiological side, it is the succession of neural tremors, variously combining into neural groups." Has the " Psychoplasm" any other or sounder basis to stand on than the argument which has now been examined? This I will not undertake to say. The conception itself must first be made clearer than appears from these two volumes of Mr. Lewes.

Before going farther I would remark, that the discussion of Mr. Lewes' Problem has carried us into psychology; we are no longer discussing a metaphysical but a psychological question; the nature of the connection between neural processes and states of consciousness is our object-matter. For his metaphysical principle of the identity of cause and effect

[1] Problems, &c. Vol. I. p. 118, 119.

is brought in by Mr. Lewes in proving a case which he supposes to fall under it, that case being the connection between nerve and consciousness. That Mr. Lewes has up to this point failed, as I have tried to show, in proving his case, makes no difference in respect of the department of knowledge to which it belongs.

But we are soon brought back again to wider ground. I have not yet mentioned all the reasons which Mr. Lewes urges in support of his view. One remains, to which great importance is apparently attached. It is that "psychological law," in accordance with which we do, as a fact, "express all objective aspects in terms of Matter and Motion, and all subjective aspects in terms of Consciousness. Motion expresses the changing positions of objects in Space —*i.e.*, redistributions of Force—and thus, metaphorically, comes to express the changes in Consciousness when these are viewed objectively."[1] Again, "The Motion which is contrasted with Feeling is strictly speaking only *one mode of Feeling contrasted with all other modes*, and made to represent the objective or physical aspect of phenomena. in preference to any other mode, because of the predominance of the organ whence it is derived,"[2] the organ, namely, of vision. And again, "Motions, apart from Vision, are as impossible as sounds apart from Hearing. Nevertheless, for the reasons previously stated, we inevitably translate all sensations into terms of Motion when viewing them objectively: as *objects* the feelings are all interpreted by the one sense which predominates in our perception of the external."[3]

---

[1] Problems, &c. Vol. II. p. 473.   [2] Same place, p. 480.
[3] Same place, p. 482.

These considerations change, by enlarging, the ground, on which the doctrine of identity of condition and consequent in the neural process stands. We are no longer asked to deal with motion in the neural process, but with motion in the whole external world, of which the neural process is but a part. The entire motion in the whole world, in connection with that in the sentient organism, is now taken as the objective aspect of the accompanying sensation in the organism. "For the sensation of colour there is required not only the rhythmic pulses of the ether, but the co-operation of the optical apparatus, together with the propagation of the stimulus to the brain, where certain changes are effected, the sum of which is this particular sensation."[1]

Now granting for argument's sake, that motion is an expression for the whole objective aspect of things, as distinguished from the subjective which is called feeling, still it by no means follows that a particular motion is the objective aspect, and not the condition, of a particular feeling. The neural process, a part of motion at large, may still have one feeling as its dependent concomitant, and another as its subjective aspect.

Two distinct questions, then, are mixed up together by Mr. Lewes; the first, as to the neural process being identical with its concomitant sensation; the second, as to motion at large being identical with feeling at large. On what ground are these questions so treated? We must look, I suppose, to the "psychological law" which has just been described,— "It is a fact that we express all objective aspects in terms of Matter and Motion, and all subjective aspects

<hr>

[1] Problems, &c. Vol. II. p. 488.

in terms of Consciousness;" and we do so for no other assigned reason but "because of the predominance of the [visual] organ."

We have waited up to the last Chapter of the book for an account of what Mr. Lewes understood by *objective* and *subjective aspects*, and how the distinction arose. He has indeed spoken before of the Subject and the Object as the factors of every psychical phenomenon;[1] he has spoken of perception being the assimilation of the Object by the Subject;[2] also of the external and the internal, the personal and the impersonal,[3] the self and the not-self;[4] but he has not, till now, told us how he would define what subjectivity and objectivity are, or what characteristics he holds to constitute them. Here is at last the answer. It is, that the predominance of visual sensibility over other modes of it leads us to interpret all kinds of sensibility in terms of motion, which is a mode of sensibility peculiar to vision; and so to interpret them is to interpret them objectively, all objective aspects, and no subjective ones, being included under this interpretation. The interpretation of phenomena in terms of motion (*matter* being supposed as common to all alike[5]) is the characteristic of their objectivity, while the leaving them as feelings not so interpreted is that of their subjectivity.

Applying this to the particular case of the neural process, the name itself shows that it is interpreted in terms of motion. It is therefore part of the objective aspect, and not of the subjective. But the sensation which accompanies it is not so in-

[1] Problems, &c. Vol. I. p. 122.   [2] The same, Vol. I. p. 189.
[3] The same, Vol. I. p. 119.   [4] The same, Vol. I. p. 194.
[5] The same, Vol. II. pp. 474. 176. &c.

ELEMENTS, ASPECTS, AND CONDITIONS. 51

BOOK III.
CH. VII.

§ 3.
Conditions of
Phenomena.

terpreted ; and therefore, remaining as feeling, it belongs to the subjective, and not to the objective, aspect.

But, according to Mr. Lewes, its concomitant sensation is the only subjective aspect which a neural process has. For he is not aware, as I have shown above, of its true subjective aspect. He sees, therefore, nothing to hinder him from abolishing the neural process *as a condition*, and taking it solely as a part of the objective aspect. He first takes its effect, its concomitant sensation, as its subjective aspect, then brings it together with the sensation under his general distinction of aspects, and the terms *condition* and *conditioned* cease to be applicable. But this whole argument fails, when it is shown, as above, that the neural process has both a subjective aspect and a concomitant sensation, wholly different from each other. For in this case we must have a theory which provides an appropriate place for both.

And now it may fairly be asked—What is the outcome of all this reasoning of Mr. Lewes ? What sort of world do we live in, when the distinction of condition and conditioned has been swallowed up in that of opposite aspects ? What are these opposite aspects which remain as our ultimate distinction ? Again I remark, that the Chapter now under discussion is not the only place where the subjective and objective aspects are spoken of, though it is the place where the final explanation of them is to be found. In this Chapter we read that "the Motion which is contrasted with Feeling is strictly speaking only *one mode of Feeling contrasted with all other modes.*"[1] Feeling, then, it is which includes both the aspects,

[1] Part of a passage already quoted : Vol. II. p. 480.

subjective and objective. There is motion which is one mode of feeling; then there are the other modes of feeling; and then there is feeling at large, which includes both, having something common to both, namely, "muscular innervation," which is "a necessary factor in every feeling."[1] It might seem from one passage, as if even this most general feeling had its correlate, namely Matter; since we read that "Matter has its indestructible Activity, and phenomena are its manifestations. But Matter to us is the Felt, and therefore all its manifestations are changes in our Feeling; &c."[2] Still Mr. Lewes does not insist upon the correlation.

The subjective and objective aspects, then, do not reach to everything. There is Feeling beyond them, which they divide, Feeling in general, the characteristic of which is "muscular innervation;" a characteristic which is common to both aspects, but not characteristic of either. This generic Feeling it is, which is Mr. Lewes' "ultimate," that in which the subjective and objective aspects are united. Their *de facto* community is expressed by referring them to one logical genus, notwithstanding their specific difference. "While the logical disparity between Object and Subject, or Motion and Feeling, is wide and irremovable, the real parity lies in their being both modes of Feeling."[3]

The latest expression, then, of Mr. Lewes' views is, that feeling, a single term, is the highest and most general term in our knowledge. The objective and subjective aspects are its modifications, and arise according as we do or do not experience the particular

---

[1] Problems, &c. Vol. II. p. 476.
[2] Same place, p. 474.　　　[3] Same place, p. 492-3.

ELEMENTS, ASPECTS, AND CONDITIONS. 53

BOOK III.
CH. VII.
———
§ 3.
Conditions of
Phenomena.

feeling of motion, or combine it with other feelings. It follows from this, that, if we had no sense of sight, and consequently no sense of motion, we should have no objective aspect of existence; what is now called subjective feeling would be all. The two aspects are thus reduced to be two species of a common genus, feeling; motion being the differentia of objectivity. The aspects are not our ultimates, but feeling; and the distinction of the aspects is reduced from the rank of a philosophical to that of a psychological distinction. Feeling, the ultimate, is taken as a something, not objectively existing but still a something,—are we to call it "Psychoplasm"?—which has two sorts of sensibility, an objective sort and a subjective. And objective existence depends upon this something, feeling, being endowed with *vision* among other senses. In my view on the contrary, the philosophical as opposed to the psychological view, feeling, so far from being the common genus of the objective and subjective aspects, is a name for the subjective aspect alone, and its obverse, the objective, aspect is existence.

Such, then, is the theory which Mr. Lewes calls Monism. The name has been chosen by himself,[1] and I think rightly. You cannot take the categories of direct thought as ultimate categories of existence, without having a single thing as ultimate genus, (in this case *feeling*); and without considering its subgenera as portions of it, or as constituents of it. However abstract or indefinite that single term may be, were it even τὸ ἕν or τὸ ὄν, it is regarded as if it were an existent and an absolute, for it is an object of direct and not of reflective perception.

But so to regard existence, as an object of direct

———
[1] *Problems, &c.* Vol. I. p. 122.

perception, involves necessarily applying to it as its highest categories the empirical and scientific categories of condition, factor, and so on. Then reflection supervenes, and you are referred from direct distinctions to reflective, and back again from reflective to direct, without end; what now appears as condition is soon seen as aspect; what now is seen as aspect appears presently as factor. Nor is there any escape from this perpetual Hegelian transformation scene, except by defining what is meant by factor, what by condition, what by element, and what by aspect. Otherwise it is a ceaseless round of binding and loosing, *setzen* and *aufheben*, asserting and retracting, without even the satisfaction of a method in the process, or a goal to which it may seem to tend. Hegelianism offers at least these; there is a distinct method in the negation of negation, and a distinct goal in the Absolute Mind.

Mr. Lewes, indeed, seems to take pleasure in retracting for its own sake; and sometimes he surprises us by his eagerness. I was surprised and a little disappointed too, after reading the sentence which speaks about the parity between Object and Subject lying in their being both modes of Feeling,[1] to find that the very next words were " I do not mean this in the idealistic sense." The logic would have been better if he had. A theory of the Absolute:—a Monistic theory of the Absolute;—a Monistic theory of the Absolute in which Feeling is sole existent;—and yet not meant in an idealistic sense,—seems to me just a little, *tant soit peu*, illogical.

You have a monistic theory of the absolute, if you make a single genus, which is a category of direct

[1] Quoted above : Vol. II. p. 492.

thought, your ultimate of all thought. You have an idealistic monism, if that ultimate is of a subjective character. The *partial* character inherent in idealism, the *direct* character inherent in monism, are the ruin of both as philosophical theories. Facts *force* us to have *two*, the two *reflective*, aspects as our ultimates of thought.

Even Mr. Lewes cannot avoid it. What prompted the "I do not mean this in the idealistic sense"? I believe he did not *mean* to mean it. His *theory* means it, not he. Thus, too, we saw above,[1] that his analysis has really furnished him with an objective correlate of Feeling, namely Matter, if only he would stoop to take it up. And, strangest of all, in making his final statement and summary, he falls at once into the assumption of reflective aspects: "Existence—the Absolute—is known to us in Feeling, which &c."[2] Feeling, the quondam ultimate, appears here but as the subjective aspect of existence or the absolute. There is, then, something beyond even Feeling? There is. And what may it be? The object of the feeling, Existence.

Perhaps it will be said, that at least Mr. Lewes has the merit of having solved, or rather dissolved, the question, hitherto held insoluble, as to the mode of connection between neural process and consciousness. He lays claim to have done so by proving the identity of the two.[3] But this question seems to me to remain precisely in its old position. For it is a question, not of metaphysic, but of psychology; to show that a neural process has a subjective aspect in metaphysic is no explanation of its having a consc-

[1] At p. 52.　　　[2] Problems, &c. Vol. II. p. 502.
[3] Same place, pp. 458. 459.

quent in psychology, notwithstanding that both its
consequent and its subjective aspect are states of
consciousness.   More than this must be shown in
order to solve the question, namely, that the consequent and the subjective aspect are one and the same
state of consciousness.   But it is just this which Mr.
Lewes has not succeeded in showing.

[NOTE. The foregoing criticism was written before
the appearance of Mr. Lewes' third volume of Problems, *The Physical Basis of Mind*, where he returns
to this question in the Problem entitled *Animal Automatism.*   He there makes his meaning much more
clear, so clear that his argument may almost serve as
its own refutation.   From § 30. of that Problem the
state of consciousness accompanying a neural process
is plainly conceived as both subjective aspect and
conditioned result of the process.   For, after describing it as aspect, Mr. Lewes proceeds, " Secondly, between the logical proposition and the physical process"
[which is his instance of the phenomenon in question]
" there is a community of causal dependence, *i.e.*, the
mode of grouping of the constituent elements, whereby this proposition, and not another, is the result
of this grouping, and not another."[1]

I have therefore nothing to alter in my argument.
For on this passage I have merely to ask—If the
logical proposition, which is the result, is also the
subjective aspect of the neural process, how do we
know *that we have* the neural process?   We should
know it *only* as logical proposition, and not as neural
process at all, if the logical proposition were its sub-

[1] Physical Basis of Mind, Prob. III. Chap. 3. p. 335.

jective aspect; it would *be* a logical proposition to consciousness, and nothing else.

The use which Mr. Lewes makes of this doctrine, in the Problem *Animal Automatism*, is to maintain against Professor Huxley, that consciousness (in the large sense of feeling or sentience) is an agent in determining the actions of the living being. It is argued to be so on the ground of its *identity* with the neural process: what the neural process does, *it* does. "Surely it seems more accurate to say that it accompanies *and* directs the working?"[1] "The feeling which *accompanies* one muscular contraction *is itself the stimulus* of the next contraction; * * * the collateral product of one movement becomes a directing factor in the succeeding movement."[2]

Now all that Mr. Lewes says on the wide and fundamental difference between vital mechanism and merely mechanic mechanism, and on the power of "selective adaptation" shown by the former, is to my mind sound and thoroughly admirable. The living organism is not a dead but a living machine, capable of acting *pro re nata*, and capable also, wherever consciousness is present, of choosing and of willing. But the question which is raised under the title "automatism" is not whether a sentient organism can act selectively, but whether the *sentience* of that organism, *taken apart*, contributes to determine the act. In other words, the question is a question, not of metaphysic, but of psychology. It is so because psychology separates its phenomena, is science and not philosophy. To answer this psychological question by a metaphysical, a philosophical, distinction, that of the two aspects, is beside the mark. The

[1] Physical Basis, &c. Chap. 7. p. 406.   [2] The same, p. 407.

psychological question, relating to sentience *by itself*, is not thereby answered. Not to mention that, as I have tried to show, the attempt to identify the metaphysical *aspects* with the psychological *condition and conditioned* entirely breaks down. Consequently, on the one hand, psychology is assured against the unwarranted intrusion of the metaphysical notion of *aspects*, and on the other hand, a result if possible still more important, metaphysic is protected from the equally unwarranted intrusion of the psychological *condition*. True, the notions of aspect and condition are defined by metaphysic, and both in this sense belong to it; but the *use* of the two notions is different; that of aspects being of use chiefly in metaphysic, that of condition and conditioned for the most part in science, of which psychology is a branch.]

§ 4. I will conclude this Chapter by indicating the exact point and exact mode of divergence between the two domains of metaphysic and psychology. This will best be done by returning to the point where Mr. Lewes has confused condition with aspect, by neglecting to observe the *two* subjective aspects which he introduces.[1]

The tactual sense and its special percepts, namely, solidity resistance and motion of solids, are the field in which this confusion arises, and in which an accurate analysis will disclose the difference which dissolves the confusion. The tactual sense and its percepts are the only case in which the three things,

[1] Above, p. 43.

ELEMENTS, ASPECTS, AND CONDITIONS. 59

BOOK III.
CH. VII.

§ 4.
Limits of Meta-
physic and
Psychology.

subjective aspect, objective aspect, and condition of both, appear to coincide. This sense is the only one which appears to be objectively (*i.e.*, in relation to other objects in consciousness), just what it is subjectively, (*i.e.*, in relation to consciousness alone). Resisting and moving solids appear to affect our consciousness precisely as they affect other solids and their attributes which are not solid. They appear to act on our nerves as solids act on solids; and the combined action of nerve and external solids appears to be accompanied by sensation in our nerve substance, just as the combined action of external solids is accompanied by the non-solid attributes of solids, *e. g.*, heat and colour.

Tactual sensation and its objects, moving and resisting solids, therefore, appear to contain, if not to be, the common summit, or common starting point, from which flow down the two streams of history or science, (which is history conceptualised), and of subjective analysis or metaphysic.

In history, or its generalisation, science, the question occurs,—In what point are cases of succession or co-existence which are *dependences* distinguished from cases of mere succession or co-existence without dependence? As, for instance, day and night, as depending on the positions of earth and sun, from day and night as succeeding, but not depending on, each other. Why do they not depend on each other? Experience shows that it is only in the case of *solids* in interaction that dependence exists, this dependence being of two kinds, 1st, dependence of latter on former states of the solids themselves; 2nd, dependence of non-solid attributes on states of solids.

All causation, all *real* conditioning (in the *common*

Book III.
Ch. VII.

§ 4.
Limits of Meta-
physic and
Psychology.

acceptation of the term *real*) is, so far as we know, a case of interaction of solids. These interacting solids condition each other, and together condition the co-existing or succeeding phenomena which are not solids ; but these latter phenomena do not re-act upon the solids which condition them, nor do they act upon each other. To give an explanation of anything in terms of motion of solids is a materialistic explanation ; and this materialism is perfectly legitimate ; but it is science, not philosophy.

But when we come to questions of ultimate nature, of subjective analysis, then solids and their interaction are one case, among many, of states of consciousness taken empirically or in the concrete, and, like all other empirical states of consciousness, are capable of a metaphysical analysis into elements. Motion of solids is a particular kind of motion. Motion is itself a particular kind of change, that change which takes place in space as well as in time. Change is, then, the first empirical thing, more ultimate than motion. Those Idealist systems which explain the genesis of the world by Change (*e.g.* Hegel's; for his negativity, *Unterschied,* and the well-spring of force which lies in these, are nothing but other names for change), are more profound than the Materialist systems which profess to be systems of philosophy. For these explain the same thing by something much more in need of analysis, namely, motion of solids. Still the Idealist systems spoken of are radically guilty of the same kind of error as the Materialistic; the error of posing as explanations of genesis at all. Subjective analysis is no scientific, no historical, explanation. It is purely philosophical.

Psychology belongs to science and not to philo-

Book III.
Ch. VII.

§ 4.
Limits of Meta-
physic and
Psychology.

sophy, and for this reason, that it deals with a particular class of *real* (in the common sense) causes or
conditions of dependence of phenomena on one
another. The effects or results may be general,—
all consciousness; but still the conditions are a particular class, nerve-motions, or, at the widest, motions
of solids among which nerve-motions must be included. Motions of solids are a particular class of
phenomena; the additional circumstance of nerve-
motions being found with the others distinguishes
the conditions contemplated by psychology from
those contemplated by mechanic, physic, chemistry,
and the lower branches of physiology.

These, then, are the two streams flowing down
from what appears to be the common summit or
source, tactual sensation and its objects. The question which this common source suggests is,—Whether the sole fact of a common *source* of philosophy
and science does not show the priority of science to
philosophy? *source* being a word of genesis and
history. Or, not insisting on the word, as being
possibly only figurative, does not the source being
common to both show that a common unity, higher
than both, exists? And if so,—then, this unity
being tactual sensation and its objects, a materialistic ontology is involved, which I take to be the
true meaning of the claim put forward by Materialism to be philosophy as well as science.

The answer to these questions must be given by
a further analysis. I have already shown that Mr.
Lewes has included two distinct things under his
subjective aspect, in the cases of sight and hearing;
the neural process (or nerve-motion) has a subjective
aspect as a complex of visual and tactual sensations,

Book III.
Ch. VII.

§ 4.
Limits of Meta-
physic and
Psychology.

and another (which I call its effect or result) as the accompanying sensation of sight or of hearing. Here we have neural processes subserving sight and hearing. But suppose the neural process is one subserving tactual sensations alone,—do not then the accompanying sensation and the subjective-aspect sensation melt into one and the same sensation, so as to leave us with the neural process as at once condition and objective aspect of a sensation which is at once its effect and its subjective aspect?

This is what I meant above, by saying that tactual sensation and its objects were the only case in which subjective aspect, objective aspect, and condition, *appear* to coincide. Do they coincide truly? It is for analysis to answer. They do not. It is a case of *dolus latet in generalibus*. The *accompanying* sensations of the neural process subserving touch are precise and various. The *subjective-aspect* sensations of the same neural process are most imperfectly known to us, so much so that we have to describe them in the same most general terms,—nerve-motion and so on,—in all cases, no matter what the particular accompanying sensations may be. True, the differences, which we have as yet no means of observing, in the nerve-motion, correspond minutely to and vary with, as we cannot but imagine, the minutest differences in the accompanying sensations. They would not be their *conditions* if they did not. But this varying with sensations is not the same thing as being the objective aspect of sensations. If we could see and feel the nerve-motions subservient to touch, even if we could see and feel them in all their variations, we should not be feeling the objects of tactual sensation which we perceive by their action.

BOOK III.
CH. VII.

§ 4.
Limits of Meta-
physic and
Psychology.

The objects which accompany their action are dif-ferent from the objects which compose their action.

It is, therefore, an error arising from insufficient analysis of general terms, the terms of matter and motion, to suppose that tactual sensation and its objects are at once the condition and the objective aspect of all other existences. It is one particular kind of motion of solids which is the condition of existence of all our states of consciousness. But this is very different from saying that this particular kind of motion of solids, or that motion of solids at large, is the objective aspect of all existence, or, in other words, is the whole of existence and co-extensive with consciousness. Our states of consciousness come to us piecemeal to some extent, and in consequence of reflection we find that one particular set of them, nerve-motions, accompanies as its condition whatever state of consciousness may come to us. But the uni-versality of this condition, universal so far as our experience of conditions goes, cannot render the con-dition co-extensive with what it conditions, still less render it an absolute source of that of which it is itself a part, namely, existence at large.

Many of our imaginations which have an emo-tional content are known to us only as representa-tions. If we could have presentations of these objects, as they are imagined, and if besides these presenta-tions were found to consist of, or depend upon, motion of solids, as our actual presentations do, then there would be more ground for supposing that motion of solids was co-extensive with existence at large. But as it is, nerve-motion is the condition of our having imagination and emotion, but it is not the condition of the *actuality* of what we imagine and emotionally

Book III.
Ch. VII.

§ 4.
Limits of Meta-
physic and
Psychology.

feel. For the presentation, or (same thing) for the actuality of this, we should probably need quite different senses from those which we actually have. But its actuality and its presentation are not impossibilities: there may be beings more richly endowed than man, or (same thing) there may be modes of consciousness greater and better than his, for which the presentation and the actuality exist. Man's mode of presentation and man's perception of the actual are not the measure of all actual perception, but only of his own. What may be actual to others is imaginable by him only as possible. It thus falls within the domain of reflection and philosophy, though beyond the domain of scientific verification. And in philosophy, inasmuch as it involves the hypothesis of new modes of sensation and a new psychology, it falls into the Constructive Branch, and does not belong to the analytical branch only, which is metaphysic.

I was the more alive to the confusion between aspects and conditions (a confusion which, as we have seen, is very far from being confined to Mr. Lewes), because it is one which I once shared, and which I found it most difficult to get rid of. Notwithstanding that I drew the distinction itself pretty clearly in one part of *Time and Space*,[1] and insisted upon it as one of the most essential in philosophy, the confusion still clung to me, and made its effects manifest in other parts of that book. I mean that I had not then succeeded in applying the distinction in detail to the cases which belong to and are governed by it. In particular, I had not seen that consciousness in all its parts belongs to the conditions *cognoscendi*, and

[1] In Part I. Chap. III. § 18. p. 144-149.

Book III.
Ch. VII.

§ 4.
Limits of Meta-
physic and
Psychology.

never to the conditions *existendi*. And I actually made use of the distinction between the subjective and objective aspects, in their character of inseparability, to facilitate the introduction of states of consciousness, especially states of *volition*, into the conditions *existendi* of phenomena, as links in the chain of causation.[1] I used, in fact, the distinction of aspects to save the theory of the conditioning efficiency of volition. But I could not have done this, if I had clearly and fully seen the difference between conditions and aspects, had clearly seen what I have now pointed out, that the nerve-motions subservient to any state of consciousness have a subjective aspect of their own, different from the state of consciousness which they subserve; that volition, being a state of consciousness, is subserved by a nerve-motion, and consequently is superfluous as a link in causation, which is carried on from one nerve-motion to the next, action and reaction taking place only within motion of solids.

This particular error I had seen through and expressly retracted in the *Theory of Practice*,[2] having then worked up to the view of a consistently materialistic psychology, as distinct from philosophy. This was in fact the step which enabled me to define the limits and relations between a genuine philosophy and a scientific psychology,—I mean, the removal of causation from consciousness, as such. Retain it there, and you must both materialise philosophy (since the only known causation is material), and make psychology illusory (since causation by consciousness is incalculable). In this limitation and definition of

[1] Time and Space, Part I. Chap. V. § 30. pp. 278-283.
[2] Theory of Practice, § 57. 2. Vol. I. p. 417-418.

Book III.
Ch. VII.

§ 4.
Limits of Meta-
physic and
Psychology.

the relations between psychology and philosophy lies whatever merit may be mine; and not of course in tardily admitting, and joining the ranks of those who had established, the truth of a scientific and materialistic psychology.

CHAPTER VIII.

·

NATURE AND HISTORY.

§ 1. THE discussion in the foregoing Chapter has led us to a point from which we can see as from an eminence the map of the whole territory of philosophy. This point is the distinction between nature and history. When clearly and distinctly seized, it introduces unity into the whole series of distinctions, and makes them into a single system of philosophy. In professing, as I do, to propound a new *system* of philosophy, it is upon the introduction of this distinction between nature and history that I rest the claim; because by it the true relation, not only between the branches of philosophy, analytic and constructive, but also between philosophy and science is made manifest. The distinction itself is not new, but well nigh as old as metaphysic itself. No one insists on it more strongly than Plato,—ἕν μέν τι γένεσιν πάντων, τὴν δὲ οὐσίαν ἕτερον ἕν. But the distinction has been neglected. All recent philosophies confound the enquiries into the two things, and consequently confound analytic philosophy with constructive. This was shown in the first Chapter. But now that we have seen the nature of the facts upon which the

Book III.
Ch. VIII.

§ 1.
Solids in mo-
tion and their
attributes.

distinction rests, we can appreciate the support which it gives to what in that Chapter was stated as a matter of observation.

Of the three things distinguished in the Chapter immediately foregoing, the two first, elements and aspects, belong to statical and analytical enquiry, and fall under the head of *Nature*. The third, conditions, belongs to dynamical and constructive enquiry, and falls under the head of *History*. The distinction between static and dynamic in philosophy was drawn, and the necessity of not mixing up what belongs to the one with what belongs to the other was insisted upon, by Aristotle, in the case of definitions. His words are :[1] Εἷς μὲν οὖν τρόπος τοῦ μὴ διὰ γνωριμωτέρων ἐστὶ τὸ διὰ τῶν ὑστέρων τὰ πρότερα δηλοῦν, καθάπερ πρότερον εἴπαμεν· ἄλλος, εἰ τοῦ ἐν ἠρεμίᾳ καὶ τοῦ ὡρισμένου διὰ τοῦ ἀορίστου καὶ τοῦ ἐν κινήσει ἀποδέδοται ὁ λόγος ἡμῖν· πρότερον γὰρ τὸ μένον καὶ τὸ ὡρισμένον τοῦ ἀορίστου καὶ ἐν κινήσει ὄντος.

Now the analysis into elements and the distinction of aspects give the *nature* of a thing as ὡρισμένον, μένον, ἐν ἠρεμίᾳ. The analysis into elements gives it as it is for consciousness alone, in simple inspection; and the distinction of aspects enables us to see whether it is taken as a mode of consciousness or as a mode of existence. The two modes of consciousness, simple (which includes primary and direct) and reflective, are both satisfied by these two enquiries; simple consciousness by the analysis into elements, and reflective by that into aspects. The two enquiries together exhaust the whole *nature* of the object-matter enquired into, whether that object-matter is the whole course of the world's history or

[1] Topica Z. 5. p. 142 a. 17.

the minutest possible portion of it: whether, again, it is the largest possible term of extension or the most delicate and complex modification of it.

But when we turn to the *conditions* of anything, we *eo ipso* separate it from other things and bring it into relation with some of those things from which it is separated. We are enquiring into its *genesis*, as a separate thing. Its conditions of existence lie outside it, and it arises in consequence of them; it is regarded as a member in a chain or series of events, of things that come to exist, or of things that come to pass. We cannot do otherwise than make our enquiry partial, when we institute an enquiry into genesis. There is one nameable object-matter which has no *prius*, namely, the universe as a whole; you cannot institute an enquiry into its genesis, because in doing so you would either fall into a contradiction of the definition of "universe," or into a contradiction of the definition of "enquiry into genesis." So also it is with the strictly universal and necessary elements of this same object-matter, namely, Time as a whole, and Feeling as a whole. So also it is with each of the two aspects of the universe. You cannot enquire into their genesis without falling into a contradiction. But you may take any portion whatever of time, or of time and space together, abstract or concrete; and you may take this either as a portion of consciousness or as a portion of existence; and enquire into its genesis without any contradiction. But then you are pre-supposing the analysis both of the thing enquired into and of the things which will be discovered as its conditions; that is, you are supposing this analysis to be derived from elsewhere than from your actual enquiry; and you are taking the thing enquired about as an object of the direct mode of consciousness only, that is, as an object separated from but in relation with other objects in consciousness or in existence.

In this way it is that the distinction between nature and history, a distinction in and of philosophy, drawn by reflection, gives us the distinction between philosophy and science; for all enquiries into the history of things, as distinguished from their nature, belong to science, and are scientific enquiries. Let us see what are the principal lines of scientific enquiries which take their origin from this point.

If we take the objective aspect of complete empirical things, we find our object-matter consisting of solids in various combinations and in various modes of motion, we are in a world of what is called Matter and Forces. All other qualities of objects, their light and darkness, their colour, their sound, their heat and cold, their taste, their odour, are attributes, are states of consciousness which arise in nerve substance on its being brought into connection with these solid moving objects. The resistance, solidity, and motion, of the objects themselves are likewise subjective in the last resort; but this group of feelings or qualities are now sundered from the rest, and set apart, in combination, to form the objects themselves, to which the remaining feelings are referred as qualities; we have separated one group of states of consciousness as "things," and referred the other states of consciousness to it as its qualities or attributes. Changes in these qualities or attributes are always explicable, even when they are not actually explained, by being referred to particular and definite changes in the motions of the solids to which they

BOOK III.
CH. VIII.
§ 1.
Solids in mo-
tion and their
attributes.

belong. These two groups of changes, namely, those in the solids themselves, and those which depend upon them in the attributes, are the object-matter of enquiry in the various physical sciences including physiology.

But these attributes are still, in science, taken objectively, abstracting from the fact of their being states of consciousness. They are referred to the solids and their movements as if belonging to them, as well as depending on them, and in this character are, to the physicist, the most valuable *evidence*, *causæ cognoscendi*, of the changes in the solids themselves. This distinction between solids and attributes of solids gave rise to the famous distinction of the primary, secondary, and secundo-primary, qualities of body. And we can now see what position this distinction holds in philosophy. It is not strictly speaking a philosophical distinction at all; but it is one which supposes bodies to exist, and then distinguishes, in that objectively-taken matter, the essential from the non-essential properties of it as corporeal matter. It is a distinction professedly philosophical but in truth physical, and therefore an encroachment on the territory of science; and therefore also, being really physical, it de-philosophises philosophy by being taken up into it.

The science of psychology begins when, after the examination of the modes of motion found in nerve substance, in connection with the modes of motion in its environment, the distinction of objective and subjective in states of consciousness is re-introduced; and this re-introduction consists in the attempt to connect the changes which take place in nerve and environment with those which take place in the

Book III.
Ch. VIII.

§ 1.
Solids in mo-
tion and their
attributes.

states of consciousness as such, which arise in
dependence on the former. And psychology is a
branch of science, and not itself philosophy, first,
because it assumes as known, and rests upon, the distinction of subjective and objective aspects; secondly,
because it brings things that belong to these two
aspects into connection with each other as *separates*;
and thirdly, because it is enquiry into the genesis or
history of the subjective part of its total and mixed
object-matter.

Psychology is, as Mr. Spencer calls it, a science
entirely *sui generis*;[1] it connects together portions of
the subjective with portions of the objective aspect
of existence, as separate things; bits of subjectivity
with bits of objectivity, both taken objectively, as we
are enabled to take them by reflection. But it is *sui
generis* not for the reasons which he assigns for its
being so, but for those which he assigns for what he
calls Æstho-physiology being so. Psychology as a
whole, including what he calls æstho-physiology, is
in that unique position which Mr. Spencer assigns to
his æstho-physiology alone, the position of a link
between the purely subjective and the purely objective departments of knowledge.

It may seem a singular result that a character
should be rightly attributable to a science, but on
grounds which justify its attribution not to that
science but to another. The reason is singular too.
Mr. Spencer admits no less than four distinct sciences,
in psychological matters, each unique in its own way.
There is first æstho-physiology, which "belongs
neither to the objective world nor to the subjective
world; but taking a term from each occupies itself

---

[1] Principles of Psychology, Vol. I. p. 141. 2nd edit.

BOOK III.
CH. VIII.

§ 1.
Solids in mo-
tion and their
attributes.

with the correlation of the two."[1] Next, there is subjective psychology, "a totally unique science, independent of and antithetically opposed to, all other sciences whatever."[2] Thirdly, there is objective psychology, which by using the element of feeling or consciousness to interpret nervo-muscular adjustments "acquires an additional, and quite exceptional, distinction."[3] And fourthly, there is psychology itself, formed of the union of the two latter, "the two forming together a double science which, as a whole, is quite *sui generis.*"[3] Now either Mr. Spencer's subjective or his objective psychology, or else his double science consisting of the two together, being thus distinguished from his æstho-physiology, must inevitably become a claimant for functions which I assign to metaphysic.

I have therefore to urge, in the present place, against this view, that it is logically impossible to sever psychology from æstho-physiology. Mr. Spencer's distinction between them is, briefly stated, this: Æstho-physiology deals with the connection between nervous changes and feelings; psychology deals, not with this connection, nor with that between nervous changes in the organism, nor yet with that between changes in the environment, but with the connection between these two connections. Thus psychology enquires how there comes to exist within the organism a relation between *a* and *b*, say a feeling of colour and a feeling of taste, which in some way or other corresponds to a relation in the environment, between A and B, say between the colour and the taste as existing in a fruit outside the organ-

[1] Principles of Psychology, Vol. I. p. 130.
[2] Same place, p. 140.   [3] Same place, p. 141.

Book III.
Ch. VIII.

§ 1.
Solids in mo-
tion and their
attributes.

ism. Æstho-physiology studies, say, the action of light on the retina and optic centres; psychology studies the connection of the action of light and its result, colour, with the action of juices and its result, taste.[1]

What logical difference, in point either of method or of object-matter, there is between these two kinds of enquiry I am at a loss to see. In both alike there are solids in motion connected with subjective states; the second enquiry is merely a more complex case of the same phenomenon as the first. The difference between them does not even rest on a difference in space-position; for changes in the environment are contemplated just as much by the first as by the second. The conditions which combine objective colours and tastes in the fruit and those which combine subjective colours and tastes in the organism consist alike in movements of solids; movements of the fruit particles there, of the nerve particles here. The connection between subjective states in the organism is just as much and no more dependent on the combination of those movements in the fruit, as the subjective states singly are on the movements singly. The true logical distinction falls not between enquiries which combine motion with feeling and enquiries which combine two or more connections of motion-and-feeling; but it falls between enquiries which combine motion with motion, and motion with feeling, on the one side, and those which combine feeling with feeling on the other:—enquiries objective, mixed, and subjective; physic, psychology, philosophy.

The third of these three divisions, the purely

[1] Principles of Psychology, Vol. I. p. 132-133.

subjective one, is that which has already occupied us in the preceding Chapters. Its partial or dynamic consideration gave us the object-matter of the Chapters on Percept and Concept, and on Contradiction and Contrariety. The two orders, perceptual and conceptual, exhaust the whole subjective aspect of existence considered as a process. The spontaneous redintegrations, which in that process are modified by volition, are not examined with respect to what processes, in nerve or environment, support or produce them, but with respect to what they themselves are and contain. In this way the perceptual images of which they consist are changed into concepts, and the conceptual order is produced out of the perceptual, without any question having been asked as to how or why the spontaneous redintegrations, which are the origin of the whole, occur in their actual order, consist of the images which they actually consist of, or exhibit the lacunæ which they actually exhibit. All these are questions for psychological enquiry, for an enquiry which can take place only after the distinction has been drawn between the motions of solids, which condition them, and the changes in attributes of solids, which are conditioned by the motions. And this is a distinction which arises in, and can only be given by means of, the process of thought or conception modifying spontaneous redintegrations. We have, it is true, in the whole purely subjective branch, a distinction of it into condition and conditioned; because the process of thought is a breaking up of a whole into parts and a recombining of those parts into other wholes. But the conditions and the conditionates of which this subjective order consists are conditions and conditionates *cognoscendi*, and not *existendi*; they are *argumenta*; reasons of knowing and the things known in consequence. What these conditions produce are not *things* but *convictions*.

When however we turn to the second of the three domains, the mixed domain of psychology, there the question is as to the conditions *existendi* of the states of consciousness which compose both the spontaneous (perceptual) and the volitional (conceptual) redintegrations. What are the conditions which are invariably and immediately followed or accompanied by perceptions, by trains of perceptions, by trains of conceptions? The states of consciousness themselves, whose conditions *existendi* are sought, are thereby *eo ipso* considered as *existents* and not merely as states of consciousness, or *evidences* of existents. True, they *are* existents, but they are existents of a peculiar kind. They are not solids in motion; they are everything else but solids in motion; for this is just the group of things which has been selected, out of existence at large, as containing all *real* conditions as they are called.

The qualities or attributes which we perceive as in visible objects (except solidity, resistance, and motion of solids) are a case or instance of the general distinction between the existents which are solids in motion and the existents which are other states of consciousness. We are familiar with this case of the general distinction, and therefore it offers us an analogy for representing to ourselves the general case; the relation between a solid object and its qualities of sound, colour, taste, odour, and warmth, is a means of interpreting the relation between solid objects in motion and all other states of conscious-

Book III.
Ch. VIII.

§ 1.
Solids in mo-
tion and their
attributes.

ness generally, whether these are feelings of sense, or intellectually apprehended images. The emotion, for instance, which we feel at the sight or thought of our home, in returning after absence, is as much a quality of that visible object, when conceived as *home*, as the colour of a peach is a quality of the peach. Both belong to the class of existents other than solids in motion. Both are conditioned by solids in motion, namely, by changes in nerve substance and its environment. The general case enables us better to understand the particular case, but the particular case, being more familiar, has first enabled us to rise to the general one; very much as, in learning geometry, we are familiar with the circle long before we are told anything about ellipse, hyperbola, or parabola, and yet we do not understand the circle until it is represented as a particular case of the ellipse, and both are seen to be sections of the cone, or curves of the second order.

We now see, not indeed why there is no reaction between states of consciousness and their physical and physiological conditions, but why it was a totally unwarranted pre-supposition to expect that there should be. It comes from considering solids in motion as the whole of existence, and consequently extending to the whole of existence the phenomenon of action and reaction which characterises them. Action and reaction take place between solids in motion, but not between solids in motion and other existents. Nor do we expect this in the familiar case, where the "other existents" are taken as existing objectively; there is no reaction of a quality on the substantive thing of which it is a quality. We do not expect the sound of a bell to react on the

Book III.
Ch. VIII.

§ 1.
Solids in mo-
tion and their
attributes.

bell. The vibrations of the bell act and react on each other, and so do the vibrations communicated by them to the air, so do the vibrations (or other motions be they what they may) communicated by those of the air to the nerve. The sound itself is something over and above, which has no reaction on the solids in motion; *its* reaction is as *conditio cognoscendi* in the purely subjective order. Not only states of consciousness taken subjectively, but also states of consciousness taken objectively, as *e.g.*, sound of a bell, colour of a peach, have no reaction upon solids; but this is not because they are states of consciousness, but because they are not themselves solid. States of consciousness must be grouped in a particular way, that is, must be formed into solids in motion, before they can either act or react; but states of consciousness not so grouped can be acted on by states of consciousness which are so grouped, that is, by solids in motion, and which are in action and reaction on each other.

States of consciousness, as such, may be conceived as floating like an aura over, or inhering like an attribute in, the nerve motions which condition them. Like the colour of a star, the shape of a statue, the odour of a flower, the sound of a harp, the taste of a fruit,—*abstractions,* as we call them popularly, from the star, the statue, the flower, the harp, the fruit,—but *really,* to us, their most essential features, constitutive of their inmost nature, their life, their soul. These are states of consciousness objectively taken, and they furnish us with an analogy for interpreting the relation of states of consciousness generally to nerve substance and nerve motion.

Book III.
Ch. VIII.

§ 2.
Psychology
actual and
generalised.

§ 2. Fixing our view upon the questions of nature and history in their utmost generality, it must be remarked, that to urge these questions in respect to possible combinations of elements other than the combination which gives solids in motion as the objective condition, on the one side, and the actually known qualities or feelings as their conditionates on the other, would be to enter upon the domains of a generalised psychology, a new world imagined by means of analogy, a world reconstructed in imagination out of the *disjecta membra* of our actually known one. We should be transcending the seen, and constructing an unseen world. This would be entering upon the Constructive Branch of philosophy.

But psychology itself, the middle or mixed science of the three which have just been distinguished, forms no part of the plan of the present work; and therefore, along with it, must also be put aside all special questions concerning generalised psychology, or the constructive branch of philosophy, to which it gives rise. For before we can profitably enter at all on this special and complicated case of the relation between nature and history, it would be necessary to examine that relation in the world, and this not only in the world as it is actually known to us, but also in abstraction from the distinction between attributes and solids. *All* qualities, those which constitute solids in motion as well as those which now appear as their attributes, must then be treated alike; that is, the peculiar position which solids in motion hold in psychology, as objective conditions *existendi*, must be abstracted from. And the questions must be put, 1st, How comes causation to be attached solely to solids in motion, in our actually

Book III.
Ch. VIII.
——
§ 2.
Psychology
actual and
generalised.

known world; and 2nd. How come visual and tactual sensations to be grouped together so as to form solids in motion at all? These questions, I need hardly repeat, are not now to be attempted.

But turning to the application of the distinction between nature and history to our actually known world, there is no lack of problems to be solved by its aid. There is, indeed, no question possible concerning either the genesis of elements, or the genesis of their combination generally; and there is no question now to be raised concerning that of the particular combination known as solids in motion. The *genesis* of that particular combination is a question for the constructive branch of philosophy. But its *history* is our immediate object; its process conceived as beginning simultaneously with its existence, and continuing along with it, but not prior in time to it. The connection in thought between its analysis and its process, its nature and its history,—this is what we have to examine. And in this there is no lack of problems to be solved.

§ 3. They fall into two main heads. I. Problems which relate more especially to the inner analysis of things or percepts, appearing in the shape of sceptical doubts as to their conceivability. Of these the most remarkable are, 1st, the Eleatic and Sceptical objections to the conceivability of motion; 2nd, the questions which relate to the limits of the world in space and time, both in the direction of extension *ad extra* and in that of division *ad intra;* which two questions give rise to the puzzles set forth by Kant in the two

BOOK III.
CH. VIII.

§ 3.
Conceivability
of Time and
Motion.

first of his four well-known Cosmological Antinomies.
II. Problems which belong more particularly to the
nature of knowledge; not things primarily, but our
cognition of them; consisting chiefly in questions as
to the *a priori* character of Axioms. These are the
two main classes of problems which can be solved by,
(or at least in connection with), the present distinction of nature and history. not indeed alone, but in
conjunction with other philosophical distinctions
which have been already explained, especially that
between presentation and representation, and that
between the perceptual and conceptual orders.

I. 1. First as to the logical impossibility or incon-
ceivability of motion in place, or of change of state,
in a solid, which is a motion, in place, of its parts.
If motion is inconceivable, all process and all history
are so too; nothing is left but nature; and, since all
the phenomenal is in motion, only the non-phe-
nomenal or noumenal remains, as that to which the
conception can apply. This is the Eleatic argument,
directly suggesting a real noumenal unchanging substrate of phenomenal changing appearances, as the
true theory of the world.

Now in the first place, in reply to this argument,
I point to Time or Duration, as one element in all
things, whether taken as phenomena or as noumena.
It cannot be got rid of. "True," the Eleatic will
reply, "it cannot be got rid of, but it lands you in
contradictions when you conceive it; only what is
statical can be conceived. Choose, then, between my
theory and Scepticism." This argument I now have
to answer. For if Time is an universal and necessary element, and yet is inconceivable without contradiction, we are living in a world where nothing is

BOOK III.
CH. VIII.
———
§ 3.
Conceivability
of Time and
Motion.

free from contradiction, where *everything* is appearance and unreality.

It is here that the percept-concept distinction is applicable. Conceiving percepts is arresting them by attention. Unless there were a flow, a motion, a duration, there would be no percepts to arrest. To conceive is to modify what is dynamic into something static. A dynamic percept becomes static not as percept but as concept. The concept unity presupposes the percept plurality; the concept sameness presupposes the percept difference; the concept rest presupposes the percept motion. These are not contradictories but contraries in sequence, following one another. The percept in each case is the thing apprehended, the concept its shape during apprehension. The thing apprehended is the condition of its shape as apprehended; the percept is the condition of the concept. Both are real, and there is no contradiction between them. To deny the percept is *a fortiori* to deny the concept, which depends upon it.

Underlying both rest and motion in empiricals is time, simple duration. Its divisibility without residuum, by points which have no duration, is the condition of its continuity in spite of its difference of content. Its infinite divisibility without residuum is the condition of its embracing all percepts and leaving no portions unconverted into statical moments. You cannot destroy motion, or change, which makes all things different; but you can reduce this motion and this change to an infinitesimal amount. You may then either fix your attention upon these infinitesimal *portions* of motion or change; or else you may imagine the thing which moves or changes to be arrested *at the point* of division, a point

Book III.
Ch. VIII.

§ 3.
Conceivability
of Time and
Motion.

which being a point of division has no duration. In either case the *rest*, which is predicated of the moving or changing thing, is imported into it by your conception, and not found in it as a percept. The appearance of contradiction between rest and motion arises from not distinguishing the different functions of consciousness, of which they are the objects; and is dissolved by proving that motion belongs to percepts as such, rest to percepts while they are being conceived, or in their character of concepts. Though predicated *de eodem* and *eodem tempore,* they are not predicated *secundum idem.* Or, if you take them in any particular instance, as predicated *de eodem* and *secundum idem,* then they are not predicated of it *eodem tempore.*

The Eleatic fallacy consists in building upon this confusion, and then attributing the *rest,* so imported by the process of conception, to the moving or changing object taken as a percept, as part of its essential equipment or analysis as a percept. Whereby *motion,* said of the same thing and at the same time, becomes a contradictory determination, and inconceivable so long as its contradictory, rest, is held fast. What reason tells us about things is thus made out to contradict what sense tells us about them. In the "flying arrow" puzzle,—the flying arrow rests,— we have the second way of taking the matter just signalised; the arrow is supposed to be arrested *at the point* of division. In "Achilles and the Tortoise" we have the first way of taking it exemplified. The infinitesimal portions of space to be traversed before Achilles overtakes the tortoise are infinite in number, and therefore the tortoise will always be ahead of Achilles, though by an ever decreasing distance. The

Book III.
Ch. VIII.

§ 3.
Conceivability
of Time and
Motion.

flying arrow is an argument aimed against motion directly, showing that, motion itself is impossible; the *Achilles* is one directed to show that, assuming motion to exist, a necessary consequence of it, namely, the outstripping of a slow by a rapid runner, is impossible.

Aristotle, who reports these arguments among others, also replies to them;[1] and his replies, though perfectly conclusive as against the arguments themselves, still leave something to be desired, in way of discovering the source whence the fallacy issues. He replies to the *flying arrow*, that time does not consist, as Zeno assumes, of indivisible moments, but is both continuous and as infinitely divisible as space, which he has proved in a former place.[2] This destroys Zeno's argument, by disproving his assumption, that a body is at rest at any moment, ἐν τῷ νῦν. But it leaves the source of the fallacy unexplained. It does not show how it is that we are almost irresistibly led to make the assumption of rest at a given moment, namely, that it is imported provisionally by the conceptual movement itself. The flying arrow is conceived as *now* in this place, *now* in the next, *now* in the next; but it cannot move in the place where it is, still less in the place where it is not yet; and this exhausts all possible places; therefore it is always, and wherever you imagine it, at rest; motion in it is impossible, and the *appearance* of it must be an illusion of the senses. The truth is, it is fixed as being now in this place, now in that, only by thought *conceiving* the process of its movement; it is always moving *from* the place which we *conceive* as that *in*

---

[1] Phys. Ausc. Lib. VI. Cap. 9. p. 239 b. 5. et seqq.

[2] Cap. 2 of the same Book, p. 232 a. 23. et seqq.

Book III.
Ch. VIII.
———
§ 3.
Conceivability
of Time and
Motion.

which it is. In Aristotle's words, time does not consist of indivisible moments; if it did, the arrow would for each of these be really at rest.

Aristotle's refutation of the *Achilles* is equally sufficient, but requires for its perfect comprehension an elucidation of another kind, as Coleridge saw when he said, that the plausibility of the *Achilles* rested " on the trick of assuming a *minimum* of time while no *minimum* is allowed to space, joined with that of exacting from *intelligibilia*, νούμενα, the conditions peculiar to objects of the senses φαινόμενα or αἰσθανόμενα."[1] The time requisite for Achilles to overtake the tortoise is finite, like the space traversed, and also, like the space, infinitely divisible. For Achilles *never* to overtake the tortoise it would be requisite that he should *never* traverse that finite space, *never* live for that finite time. But Zeno must grant, says Aristotle, that a finite space is traversable in a finite time,—εἴπερ δώσει διεξιέναι τὴν πεπερασμένην. He must grant this, and why? Because he assumes motion to exist, though only for the purpose of drawing an absurd conclusion from it. Zeno's fallacy is, that his argument really requires the opposite assumption. But why is this opposite assumption absurd? Because it implies that we live and move in a world of pure mathematical abstraction, and not in a world of sense and empirical objects. Our sense perceptions are only of the finite, of what is small relatively to the infinitely great, and great relatively to what is infinitely small. This is the absurdity which Coleridge signalises, of applying to the world of sense what is true only of the world of mathematical abstraction, and in that world only when the same logic

[1] The Friend, Vol. III. p. 92. ed. 1837.

Book III.
Ch. VIII.

§ 3.
Conceivability
of Time and
Motion.

is applied to both elements of motion, to the time as well as to the space. This too is the latent force of Aristotle's reply, namely, that a finite space is traversable in a finite time (εἴπερ ὅψει κ.τ.λ.); for it brings us back to actual presentative experience, as the only valid foundation of purely representative reasoning.[1]

Perception is always of a concrete or empirical, conception may be of an abstract, feature; but this does not make that abstract feature contradict the other parts of the concrete from which it is abstracted. That rest which is a feature *in percepts* is always a relative rest; the thing perceived at rest is so perceived in relation to something else in motion. The thing is in one respect in motion, in another at rest; to take this rest as an abstract feature, and conceive the thing as at rest in consequence, is no contradiction to its perceived motion in other respects.

Time is not, strictly speaking, "the measure of motion;" it is an element in motion, and therefore a condition of its measurement. Imagine a thing entirely devoid of movement and of change of any sort; you do not strip it by so doing of duration; its want of change requires the conception of its endurance without change, in order to be understood. You cannot conceive *absence* of change without *presence* of time, as its condition. Endeavour now to strip the thing, so conceived as unchanging, of its duration in time;—what happens? In conceiving it as also without duration you annihilate the thing, you

---

[1] Aristotle seems to have been on this track himself in a later examination of a kindred fallacy to this, in Lib. VIII. cap. 8. p. 263 a. 1 et seqq., where he applies the distinction of δύναμις and ἐντελέχεια to divisions of time, space, and motion.

Book III.
Ch. VIII.
§ 3.
Conceivability
of Time and
Motion.

conceive it as without existence. Change and motion therefore presuppose time; but time does not
presuppose change or motion. This is what is meant
by the phrase an "eternal *now;*" an existence without change or motion, enduring in time, but with
nothing to mark that time into lengths ; or, in other
words, an existence in an undivided *continuum* of
duration. Eternity is not the contrary of time, but
is time conceived as an undivided, unlimited, continuous duration.

When time is taken as the measure of motion or
of change generally, it is already supposed to be itself
divided, and a portion of it taken as an unit of measurement. We can imagine such units of time as
ideally and exactly equal. This is Newton's conception of "absolute time," which according to him "*æquabiliter fluit.*"[1] It then becomes the ideal standard of
measurement of all concrete motion and change; and
one which cannot itself be *tested,* for it is an ideal.
And it is this equable flow of time which Newton
made the basis of his doctrine of Fluxions, with its
method the Infinitesimal Calculus; equably flowing
time was the variable to which every other change
was referred, as a function of it.[2] In respect to such
puzzles as the Eleatic "flying arrow at rest," what
metaphysic does is to show the source and nature of
the mistake which attributes rest to the arrow at

---

[1] Principia. Scholium to Def. VIII. 3rd ed. p. 6. " Tempus
absolutum, verum, et mathematicum, in se et natura sua sine rela-
tione ad externum quodvis, æquabiliter fluit, alioque nomine dicitur
duratio : Relativum, apparens, et vulgare est sensibilis et externa
quævis durationis per motum mensura (seu accurata seu inæquabilis)
qua vulgus vice veri temporis utitur; ut hora, dies, mensis, annus."

[2] Cournot. De l'Enchaînement des Idées fondamentales dans les
sciences et dans l'histoire. Livre I. Ch. VI. § 55.

Book III.
Ch. VIII.
─────
§ 3.
Conceivability
of Time and
Motion.

any moment of its course; and what mathematic does, by means of the Calculus, is to show what velocity it has at any moment. Between them they give the complete reply to the fallacy.

§ 4. 2. Coming now to the second set of problems under the first head, those puzzles exhibited by Kant as his two first Antinomies, their connection with the distinction of nature and history may be made apparent as follows. If the world is finite in time, or in space, (1st Antinomy, *Thesis*), it will require an extra-mundane cause to account for its coming into existence; its genesis must be accounted for. If it is finite in point of division into ultimate simple parts or atoms, (2nd Antinomy, *Thesis*), then these atoms are one cause of its genesis, and their combination requires another cause to account for it.

But this view of the matter, founded on the *Theses* of the two Antinomies, contradicts the distinction of nature and history as established above, for it compels us to enquire into the genesis of the world as a whole, whereas it has been shown that only its history, supposed as beginning simultaneously with its existence, can be an object of enquiry. At the same time it appears, from Kant, that this thetical view is opposed by an antithetical one, maintaining the extension and divisibility of the world *ad infinitum* in both directions; and that this antithetical view has grounds of equal apparent validity in its favour. If the *theses* of the two Antinomies in question were true, our distinction of nature and history would fall; if the *antitheses* were true, it would stand.

BOOK III.
CH. VIII.

§ 4.
Infinity of
Time and
Space.

We are forced, then, to an examination of the two Antinomies. Kant's solution of them is not complete. It consists in showing that the major premiss, in the reasonings by which the Thesis and Antithesis are severally supported, is in each case a proposition relating to "Things-in-themselves," while the minor premisses in those reasonings relate to phenomena, and that therefore both Thesis and Antithesis are fallacious, are cases of the fallacy known as *sophisma figuræ dictionis*. For in the major premiss, he says, we assume a merely *logical* law, the necessity of having *complete* premisses for any given conclusion,—which is true as a logical law, but does not involve the necessity that such complete premisses should be actually found for the conclusion required. We then, he says, take this logical law as if it were a law, or general fact, of real phenomena, and subsume the minor premiss under it. But when we come to examine both Thesis and Antithesis, the conclusions of these two subsumtions, we find them contradict each other, because the phenomena, subsumed severally by the two minor premisses, do not alone exhaust the totality required by the major premiss. These phenomena give us merely an endless *Regressus* towards, but never a coming up with, the required totality of conditions. The truth according to Kant is, that we are not warranted in assuming that (logical) Totality as applicable to phenomena; if we do so, we are treating those phenomena as if they were "Things-in-themselves," because we assume them to be the objects contemplated by the merely logical law of Totality.[1]

[1] Kritik der R.V. Der Antinomie d. R.V. 7ter Abschnitt. Kritische Entscheidung, &c. pp. 376-379 of Hartenstein's edit. 1853.

Book III.
Ch. VIII.

§ 4.
Infinity of
Time and
Space.

This solution of Kant's is incomplete, not for the reason given by Hegel, but because it does not explain all cases of the apparent Antinomy. It leaves unexplained those in which no propositions relating to "Things-in-themselves," no proposition stating purely logical laws. *e.g.*, of Totality, as general facts of nature, are introduced. Kant was full of his new doctrine of the "Thing-in-itself;" and here he uses it to explain, not only how the Antinomies are to be solved, but also how they arise. He uses it to *state* the difficulty as well as to remove it. He makes the problem for himself to solve. For which reason his explanation breaks down, as I shall presently show, when he comes to apply it to remove the difficulty wholly, or in its general shape.

Hegel has two long and admirable notes in his *Logik*, in which he criticises Kant's Antinomies, and the two first of them particularly, at some length. In the first of these two passages, which is devoted principally to Kant's second Antinomy, he says:[1] "The critical solution by means of the so-called transcendental ideality of the world of perception has no other result than this,—it makes the so-called Antinomy something *subjective;* whereby it [the Antinomy] continues to present the same illusion (*Schein*), that is, it remains just as unsolved as before. Its true solution can consist only in this, that two determinations (*Bestimmungen*), while opposed to each other and necessary to one and the same concept, cannot be valid in their one-sidedness, each for itself, but that they have their truth only in their resolution (*Aufgehobensein*), in

[1] Wissenschaft der Logik. 1ᵗᵉʳ Theil. 2ᵗᵉʳ Abschnitt. 1 Kap.
A. Anmerkung 2. Werke, Vol. III. p. 210.

the unity of the concept into which they are resolved."

Book III.
Ch. VIII.

§ 4.
Infinity of
Time and
Space.

In the second passage, which treats more particularly of the first Antinomy, Hegel says:[1] "The solution of this Antinomy, like that of the former, is
transcendental, that is, consists in maintaining the
ideality of space and time as forms of intuition, in
the sense that the world *in itself* is not in contradiction with itself, nor self-destructive; but only
consciousness, in its intuitions and in referring its
intuitions to understanding and reason, is of self-
contradictory nature. It is showing too great tenderness for the world to shift its contradiction on to the
shoulders of mind and reason, there to lie unsolved.
The mind in fact it is, which is strong enough to
endure the contradiction, but it is the mind too which
has the means of solving it. The so-called world, on
the contrary, (be it objective and real world, or subjective intuition of transcendental idealism, sensation thought by means of categories), for the same
reason never at any point dispenses with contradiction, yet cannot endure it, and on that account is the
prey of origination and destruction."

We have now before us the doctrines of these
two great masters, Kant and Hegel, in respect of
these two Antinomies. Kant wishes to remove the
apparent contradiction, Hegel to retain it. With
Kant it is a mistake *ab initio;* with Hegel it is an
essential feature of existence, only to be solved by
being preserved as a feature and embalmed in the
amber of the Absolute Idea. Hegel saw very well
that Kant's solution was not sufficient; he also

[1] The same. 2te Kap. C. Anmerkung 2. Werke, Vol. III.
p. 268.

Book III.
Ch. VIII.

§ 4.
Infinity of
Time and
Space.

thought that no solution could be. but that contradiction was essential to all existence. was the motor
of the Absolute Idea, the life of the Absolute Mind.
We shall see, however. that neither was Hegel, after
all, so far removed from Kant, nor Kant from the
truth, as Hegel imagined. They both had their
mental eye on phenomena and saw them straight,
not distorted at least by any media but those of their
own intelligence. Nor is Hegel's quiet irony, his
"*too great tenderness for the world*," unfriendly. After
all he believed himself to be developing Kant's principles; it was Kant's falling short of a solution that
caused *him* to acquiesce in contradiction. We have
the same phenomenon repeated which has been before
observed; each philosopher *does* what he professes
not to do, and *does not* what he professes to do. Kant
professes to solve but really leaves unsolved, Hegel
professes to accept but really solves, the contradiction. At least his two "moments," continuity and
discretion, contain the solution, though not, as Hegel
would have it, by union in a higher concept. A
higher concept is a mere return to the moment of
discretion. The alternation of limit-setting and limit-
overleaping, the *process* of doing so, is that which
combines the two "moments," not the concept which
results from that process. There is no higher concept
which results from the process; none into which the
continuous and the discrete. the infinite and the
finite, can be resolved; because the concepts are
already known of which they are determinations.
They are determinations of time and of space. Without these they would be meaningless; from these
they come. and to these they belong; and these are
as inexhaustible as the process of alternation. What

NATURE AND HISTORY. 93

Book III.
Ch. VIII.

§ 4.
Infinity of
Time and
Space.

is this process, however, but Kant's "*endless regressus,*" the regress towards totality of conditions in phenomena as such, and not as "Things-in-themselves"? The movement of thought by means of contradiction is not, as Hegel would have it, an union of contradictories, but a succession, a nexus by sequence, between contraries. Kant describes the lock and seeks for the key. Hegel brings the key and double locks the door. Kant is saved from Hegel's error by his clear perception of the difference between contradictories and contraries;—

"ἦ μὰν ἀμφοτέροισιν ὁμὸν γένος ἠδ' ἴα πάτρη,
ἀλλὰ Ζεὺς πρότερος γεγόνει καὶ πλείονα ᾔδη."

But it is high time to turn to Kant's general solution, that in which, without specially referring to the logical form of the Antinomies, he attempts to show that the dispute is literally about nothing at all (*dass sie um Nichts streiten*), and see how and why it breaks down, how completely the "Thing-in-itself" fails to account for the existence of the dispute, as well as failing to settle it. In the course of this general explanation we read :[1] "If we look upon the two propositions 'The world is in quantity infinite,' —'The world is in quantity finite,' as contradictories, then we are assuming that the world (the whole series of phenomena) is a Thing-in-itself. For it remains, whether I deny (*aufheben*) the infinite or the finite Regressus in the series of its phenomena. But if I do away with this assumption, or transcendental illusion (*Schein*), and deny that it is a Thing-in-itself, then the contradictory opposition of the two assertions changes into a merely dialectical one" [an

[1] Kritik der R.V. as above, p. 380. Hartenstein, 1853.

Book III.
Ch. VIII.

§ 4.
Infinity of
Time and
Space.

opposition where both may be false], "and because
the world does not exist at all in itself (independently
of the regressive series of my perceptions), therefore
it exists neither as a whole infinite in itself, nor as a
whole finite in itself. It is to be found only in the
empirical Regressus of the series of phenomena, and
not at all for itself. Hence, if this series is at every
moment (*jederzeit*) conditioned, then it is never
wholly given, and the world is therefore no unconditioned Whole, therefore also does not exist as such,
either with infinite or with finite quantity."

Kant's reasoning I take to be this: If I take
Thesis and Antithesis as expressing contradictory
determinations, I assume the world to exist as having
a definite quantity, a Whole, a Thing-in-itself, and
then the one must be true, the other false. But if
the world has no definite quantity, is not a Whole,
is not a Thing-in-itself, then the Thesis and Antithesis
may both be false, as having no subject-matter to
apply to. Now the world is *not* a Thing-in-itself,
has not a definite quantity, is not a Whole; for it
exists only in the Regressus of the series of phenomena. Therefore both the Thesis and Antithesis
may be false, and the contradiction between them is
solved.

Three things are to be remarked in this solution.
First, the empirical regressus in the series of phenomena, which constitutes the phenomenal world, is
precisely what is maintained by the Antitheses of the
two first Antinomies. "The world has no beginning
and no limits in space, but is infinite in regard to
time as well as in regard to space;"[1] and "No compounded thing in the world consists of simple parts,

[1] As above, p. 329. Hartenstein, 1853.

Book III.
Ch. VIII.

§ 4.
Infinity of
Time and
Space.

and nothing simple exists anywhere in it."[1] Kant's empirical regress and his two Antitheses express one and the same thing. They also contain both the moments of continuity and discretion, of setting a limit and overleaping it. There is no one-sidedness, as Hegel objects, in the Antitheses. There may be in the Theses, but that is another matter. The plain fact which both the empirical regress and the Antitheses build upon and express is, that it is not possible to imagine a limit drawn anywhere, in time or in space, without imagining time, or space, existing on the far side of that limit. You draw a limit by attention in reasoning; this is to turn a percept into a concept;—well, beyond that limit is a percept again. If you want to know what that further percept is, you must limit it, draw another line, turn it into a concept; this concept will have an unlimited percept beyond it, and so on;—Kant's empirical regress, and Kant's infinite time and space of the Antitheses.

Secondly I remark, that the Theses of the two Antinomies express the determination of time and space contradictory to this; they assert that there is a final limit, a final concept, beyond which there is no percept, no space, no time. In division *ad intra* we come to atoms, finite, indivisible; in extension *ad extra*, to a finite magnitude in space or in time, beyond which is no space, and no time. We have nothing at all beyond those limits. While the Antitheses and the empirical regress stand or fall together, the Theses stand or fall alone. But, since the empirical regress is another name for the phenomena as they present themselves actually to consciousness, it

[1] As above, p. 329. Hartenstein, 1853.

Book III.
Ch. VIII.

§ 4.
Infinity of
Time and
Space.

is practically impossible for the mind to acquiesce in the finality of the Theses. It *must* transcend the limits it professes to have ascertained, and gives free scope to its fancies in depicting the *supernatural* world. Those who love their fancies most are most sure to insist on the Theses of these Antinomies, for they assure free play to the fancy beyond the limits which they assign.

The third remark I have to make is to the effect, that here the Thing-in-itself totally breaks down as a solution of the Antinomies. In order to permit both Thesis and Antithesis to be false, the thing of which they are said must be assumed or shown, either not to exist at all, or not to exist as a thing capable of the determinations in question. It is not enough to say,—'these determinations are said of the world, and the world is not a Thing-in-itself, therefore they may be false.' You must say 'and the world is not capable of either of those determinations.' Neither is it enough, (indeed it is not permissible), to say 'and the world has no definite quantity, is not a Whole;' for this is to assume the truth of the Antitheses, thus assuming the very point in question. In order to get rid of the apparent contradiction, it is requisite to show that the contradictories are said of a world which either does not exist at all, or is not capable of either of the contradictory determinations. But this Kant never shows.

We mean by the world, according to Kant in the passage cited, something found "in the empirical regressus of the series of phenomena." This series of phenomena itself is the empirical world: and we put the question—Is this series finite or infinite? The "Thing-in-itself" has nothing whatever to do

Book III.
Ch. VIII.

§ 4.
Infinity of
Time and
Space.

with the answer to this question, which is wholly phenomenal. The series, the regressus, is clearly capable of quantitative determination; the question, —finite or infinite,—though we may not be able to decide it, does not fall to the ground for lack of a subject-matter to apply to. It is a distinct issue between contradictory determinations of the same thing, at the same time, and in the same respect. Which of the two is true is a question for analysis; and the analysis of percept and concept, of contradiction and contrariety, in previous Chapters, can leave little doubt on this point. The regressus is infinite, the Antithesis in both Antinomies is true, the Thesis false. In other words, the Antithesis is the true determination of the regressus.

The source of Kant's failure to solve the Antinomies by his Thing-in-itself was of a nature to blind him to the fact that he had failed. What was this source? What caused him to overlook the *general* case, where the "major premiss" is not necessarily propounded as relating to Things-in-themselves, is no mere logical law of Totality? It is this. He missed the real cause of the illusion of the Thing-in-itself, thinking that raising Categories to the rank of Phenomena was the cause. The real cause of the illusion, as I have shown in Chapter III., is the looking at existence as an object of direct, instead of reflective, perception. By so doing you make it an absolute, you treat it in and for itself, no matter that its content is phenomenal. But wherever a content is phenomenal, it is capable of quantitative determination.

What Kant should have said, then, in solution of the Antinomies, (supposing him bent on bringing in

Book III.
Ch. VIII.

§ 1.
Infinity of
Time and
Space.

"Things-in-themselves" at all), was this: The major premiss treats the phenomena as an object of direct perception, and not as an object of primary or reflective; and by so doing, since every object of direct perception is a finite thing, virtually *assumes* the *Theses* of the two Antinomies, an assumption for which there is no warrant. The Theses therefore involve the fallacy of *petitio principii.*

But treat them as a series of primary percepts, which they truly are; that is, treat them as an object of reflection, capable of quantitative determination, but not having that determination already known; and you will find no other quantitative determination of them possible but that which is expressed by the Antitheses. As phenomena simply, they come before us only in Kant's empirical regress; but since this regress involves a perpetual alternation of the continuous and the discrete, the Antitheses, which express the fact that, wherever you divide the continuous, you have the continuous again beyond it, are the only possible determinations of its quantity.

Kant on the contrary thought, that the world did not exist as an object *at all*, until determined by some Category. Until it was so determined, there was literally Nothing to strive about. But, since all determination by the Categories is limitation, so soon as anything was determined, and there existed something to strive about, the question *finite or infinite* was already beforehand decided in favour of *finite.* For the alternative was—either no world or a finite one. Not the Antinomies, then, but their Theses it is, to which Kant's Thing-in-itself gave birth. For what is Kant's Thing-in-itself? Empty Categories assumed as real phenomena. Now, if you will have *no* world

Book III.
Ch. VIII.

§ 4.
Infinity of
Time and
Space.

but what is thought by the Categories, you are *eo ipso* treating the world as a Thing-in-itself; the very process which Kant denounces as the source of fallacy. In other words, adopt Kant's view of what the Thing-in-itself is, and you cannot help treating the world as a Thing-in-itself. To do this is to assume the truth of the *Theses* of the two Antinomies. But it is in no wise to introduce the Antinomies themselves, except in so far as the facts of the case inevitably array themselves against a theory at once arbitrary and inconsistent.

§ 5. II. The second kind of problems mentioned above, that relating to the nature and functions of Axioms, has now to be treated. It is most convenient to treat it in connection with the distinction of nature and history, because the form in which the question is usually raised involves a separation between the series of subjective states, which are knowledge, and the series of objective phenomena, which are matter in motion; and then the question is put. How arises the ultimate nexus between them, how in the last resort do we know that our knowledge truly represents *things?* I therefore treat the question under the distinction of nature and history, which explains the character of the supposed separation, in order to satisfy the demands of the problem as usually presented.

It is generally agreed, that all reasoned knowledge, *i.e.*, knowledge considered as arising out of and having behind it certain rude material, in shape of percepts, as its *data*,—that all reasoned knowledge

Book III.
Ch. VIII.

§ 5.
Postulates,
Axioms, and
Hypotheses.

begins either from axioms. or else from provisional hypotheses in their stead. One school of thought holds strictly to the latter alternative : they hold that we begin by framing hypotheses. and that these when verified by experience become truths which are a firm basis for further hypotheses, which again when verified lead to further truths, until at last our first obtained truths are so completely verified by experience of all kinds as to be fairly called axiomatic.

Another school takes the other basis, and holds that we begin originally with axioms, that we then use hypotheses only to discover particular cases within the axioms, and that the whole structure of discovered knowledge is really no more than the verification and exemplification of the originally known axioms. This school, in short, holds that axioms are certain from the very first ; the other, that their certainty is acquired at last through verification of hypothesis by experience.

On these views of the case the following difficulties arise. If hypothesis is the ultimate source, as the first school holds, there is really no axiomatic knowledge at all, for the axioms at which we arrive at last are simply records of matters of fact. owing their apparent necessity to long uncontradicted experience, confirmed by habit and enforced by heredity ; our knowledge has no element of a *necessary* character in it. but is rather to be called *habitual belief.* On the other hand, if axioms are the ultimate source, we have certainty in knowledge, but no guarantee that this certainty is truly ascribed to facts ; for the axioms may be arbitrarily imposed on facts by the constitution of our own minds.

NATURE AND HISTORY. 101

BOOK III.
CH. VIII.
——
§ 5.
Postulates.
Axioms, and
Hypotheses.

These are the two opposite views of the genesis of knowledge which may be said to divide the learned world between them. They both alike issue in a separation between knowledge and fact, because both alike begin with assuming a separation between object and subject. The one issues in a combination of facts which is not really knowledge, having no strict *necessity*, the other in a connected and necessary knowledge which is not a knowledge *of facts*.

The question is, whether there is not a way out of this double inconvenience? May there not be a source of error in separating object and subject? And is not this separation involved in the assumption, either that we begin with hypotheses alone, or that we begin with axioms alone? I apprehend the truth to be, that we begin to reason, for acquiring knowledge, by framing hypotheses based on axioms; that is, we begin with axiom and hypothesis together. This is what I hold to be the philosophical position on the question; and this is what I am going to make good. But first a little more illustration of the non-philosophical positions. And I select one which leads directly to the point which I wish to establish. I take it from Mr. Spencer's Principles of Psychology.

We there find the statement, that neither the *dictum de omni et nullo*, nor Mr. Mill's 'mark of a mark' axiom, "nor indeed any axiom which it is possible to frame, can, I think, be rightly held capable of expressing the ratiocinative act."[1] Then we read: "Axioms can belong only to the subject-matter about which we reason, and not to reason itself—imply cases in which an objective uniformity determines a

[1] Principles of Psychology, Vol. II. § 303. p. 93.

Book III.
Ch. VIII.
—
§ 5.
Postulates,
Axioms, and
Hypotheses.

subjective uniformity; and all these subjective uniformities can no more be reduced to one than the objective ones can."[1] Then again: "* * there will be some *universal* necessity of correlation — some axiom. Such an axiom is therefore to be accepted as expressing absolute dependencies in the *non-ego*, which imply answering absolute dependencies in the *ego*—not, however, absolute dependencies in the *ego* that are recognized as such in reasoning.

"The utmost that any analysis of reason can effect is to disclose the *act of consciousness* through which these and all other mediately known truths are discerned; and this we have in the inward perception of likeness or unlikeness of relations. But a truth of this kind does not admit of axiomatic expression, because the universal *process* of rational intelligence cannot become solidified into any single *product* of rational intelligence."[2]

There is somewhat of an *a priori* assertorical air about this last "because," which one would like to see fortified by another. Meantime the Postulates of Logic are a sufficient answer to Mr. Spencer's denials in the passages just quoted. They express the "ratiocinative act," the act which modifies percepts into concepts, combines concepts into judgments, and judgments into syllogisms. It is the same act, in matter of different degrees of complexity. It is true that they do not express that act as *pure* act, (nor indeed does it so *exist*): but they express it as an act, abstracted from every particular matter with which it is involved; and this they do by means of the symbolic forms of expression, A and not-A. The extreme

[1] Principles of Psychology, Vol. II. § 303. p. 91.
[2] The same, p. 94-5.

of generality in the expression is the extreme of abstraction in the thing expressed.

Book III.
Ch. VIII.

§ 5.
Postulates,
Axioms, and
Hypotheses.

Now I maintain that the Postulates of Logic,
taken together, make one axiom and only one; an
axiom of which all other axioms are cases, cases of
greater complexity in point of content. All these
might also be called postulates as well as axioms;
for—postulate, axiom,—where is the difference, or is
there any? There is, and it is this. A dictum is
called a postulate in relation to the volition of the
reasoner, and an axiom in relation to the phenomena
reasoned of. But in reasoning, truth of fact being
there the purpose aimed at, you cannot make a postulate unsupported by *fact*, for it would have no legiti-
mation; nor can you assert an axiom without being
warranted in claiming assent to it, in making a *postulate* of it, in case of argument. Postulate and Axiom
are opposite *aspects* of each other; not respectively
subjective and objective aspects, but aspects turned
respectively towards volition and towards phenomena,
the object-matter of that volition. Both these aspects
together constitute the subjective aspect of Nature
itself, or have Nature as their object-matter.

The phenomenal or axiomatic aspect of the postulates of logic is the most general axiom in all knowledge. It is that which asserts the *Uniformity of
Nature*. If there were no uniformity in nature, there
could be no postulates of logic; if that uniformity
were not universal and without exception, the postulates could not be universally and necessarily true.
For while we asserted A, in the subject of the proposition of identity, in A is A, it would, or at any
rate might, be changing into not-A, while we made
the assertion of its identity, and the postulate would

Book III.
Ch. VIII.

§ 5.
Postulates,
Axioms, and
Hypotheses.

be falsified. It would not be true as a postulate, though it might happen to be true, in a particular case, as a matter of fact.

The identification of the postulates of logic with the axiom of the uniformity of nature justifies the position with which I set out, that in the beginning of knowledge we begin with axiom and hypothesis together. For we can frame no hypothesis at all but by the aid of the postulates of logic, which now turn out to be an axiom. This is clear, because every concrete hypothesis involves judgment; and the postulates of logic are thus, prior to verification by experience, the first hypothesis, notwithstanding that they are also *postulates*, that is, the *only* hypothesis which we have it in our power to make. If then we take reasoning in pursuit of knowledge to begin with hypothesis, we find that in order to frame that hypothesis the postulates are requisite, and, if the postulates, then also the axiom of uniformity of nature.

The school, then, which holds that reasoning does not begin, but only ends, with axioms, is clearly wrong, at least in the sweeping range which it often gives to its assertions of the doctrine. But is the opposite school, that which holds that reasoning begins with axioms and not with hypotheses, in the right? By no means. We do not begin our reasoning in pursuit of knowledge by the postulates alone, nor by the axiom of uniformity alone. We begin by an hypothesis in which the postulates and the axiom are involved. We say— " What is that?"[1] Then we frame an hypothesis. There is no axiom laid down, as firm ground to begin with, as major premiss of a

[1] See above in Chapter V. Vol. I. p. 292.

BOOK III.
CH. VIII.

§ 5.
Postulates,
Axioms, and
Hypotheses.

deduction; *whatever* our hypothesis may be, it proceeds on the assumption of the uniformity of nature as warranting its validity simply as a reasoning process. This is the axiom upon which all hypotheses are based, and necessarily based because they are processes of reasoning moving by the postulates of logic.

But are the Postulates of Logic universally and necessarily true?—Aye, that is the great question. They may be requisite for all reasoning, and consequently for all science; for what Kant called cognition, *Erkenntniss*, and experience, *Erfahrung*, as well as for thought, *Denken*. But what shows that science is a case of necessity? *Je n'en vois pas la nécessité*, says the Sceptic; your *reasoning* is all in the air, fancy *in nubibus*. We cannot, then, *assume* that science must be true; it may be all a cunningly wrought dream; true indeed for us, while it lasts; but *this* sort of truth is fiction. The scientist is satisfied with this sort of truth, but not the philosopher. And the scientist very justly; for why should he trouble his head with what the sceptic or the theologian may be thinking?

Now it is not proving the truth of the Postulates of Logic, merely to show that they are axiomatic as well as volitional, or even that they are objective as well as subjective; or, in other words, merely to take the question as the philosopher takes it, and not as the scientist. This indeed is essential; but it is essential as a preliminary only. The same question, virtually the same, recurs again on this higher platform. It must be proved as a philosophical question. In what remains of the present Chapter, I shall consider the generalities of this great question, along

Book III.
Ch. VIII.

§ 5.
Postulates.
Axioms, and
Hypotheses.

with some cases of particular axioms. The following Chapter will be devoted to this question alone.

There are other shoals and quicksands in this question, besides that of separating subject and object, which we have just escaped. There are those arising from the fact of the flux of phenomena, the Heracleitean πάντα ῥεῖ. For if there is strictly and literally *no stability*, there can be *no knowledge*, for knowledge is an established relation between phenomena. Nor can we escape by saying that the flux is true of the percepts, the differents in the perceptual order, while identity and uniformity are true of the concepts into which these percepts are modified in the conceptual order. For then either the conceptual order, or else the perceptual, would be *in nubibus*, one would not be a true representation or aspect of the other, or at least could not be known to be so. The two orders perceptual and conceptual would be torn asunder, just as, before, the objective and subjective aspects were torn asunder by the scientist theories.

It requires to be shown, not indeed that there is any prior truth or truths in the conceptual order from which the Postulates may be inferred, which would be absurd, since they are prior to *all* inference, but that we have in the conceptual order the very same content as in the perceptual, notwithstanding its perpetual flux. The *identity* of the two orders, in spite of their difference, must be made clear. We may indeed picture the perceptual order as an ever-changing Proteus, the conceptual order as the noose that binds him, and compels him to give up his secrets. But it must be shown that they are *his*, and not those of the conceptual order, or in other words, that the change to the conceptual order, initiated and governed

BOOK III.
CH. VIII.

§ 5.
Postulates,
Axioms, and
Hypotheses.

by the Postulates, is purely formal, and introduces no new disturbing element of its own into the perceptual phenomena, except that change of form.

Supposing the questions arising from the mere fact of the flux of phenomena in perception to be settled, there remains another set of perplexities which come into notice in the application of the postulates to percepts. Suppose we can arrest the flux, bind the Proteus, for a moment; suppose we have classified and predicted and verified; have introduced permanent relations between phenomena; have acquired knowledge on which we can venture to depend with certainty;—still this has its limits. We lay down our postulates and our axiom as true without restriction, without exception. And we have succeeded and do succeed in making out much acknowledged fact and truth by this method. But Proteus may not be fully tamed for all that. He may be trusted perhaps so far as we can see him, and no farther.

Dropping the figure,—the postulates and the axiom are laid down as possessed of much greater validity and more necessary truth than it can be verified by experience that they possess. Owing to the defect of our powers of sense, even aided by the finest instruments, and the most delicate methods, we can never verify exactly whether there is or is not strict identity between two phenomena. There must always be an interval between our powers of sense perception and the broad statements which we make about identity, equality, infinite divisibility, and so on; an interval for which our statements can never be verified by observation. But precisely at this point where observation ceases, it is imaginable that

Book III.
Ch. VIII.

§ 5.
Postulates,
Axioms, and
Hypotheses.

chaos may begin, or creation *ex nihilo* may take place. In short, the impossibility of verifying by presentations of sense impresses an hypothetical and ideal character even on the postulates and the axiom; so that it may be argued with considerable plausibility, that we begin after all with hypotheses which are not axioms; for, as it turns out, our supposed great axiom is itself not axiomatic but hypothetical.

§ 6. Questions like those which have been now stated, namely, 1st, that arising from separation of the subjective and objective aspects, 2nd, that from the fact of percepts being in perpetual flux, and 3rd, that from the impossibility of verifying axioms by presentations of sense, require a more thorough examination of the whole subject than can be given at the end of a chapter devoted to so much besides. Reserving, then, this examination for the following Chapter, I proceed to some few remarks on the subject of axioms in general.

There is another axiom closely akin to that of the uniformity of nature, indeed another aspect of it, which equally with it flows from, or rather is, the phenomenal aspect of the postulates. It is known as the Uniformity in *the Course* of Nature. The difference of the two axioms is, that one envisages single percepts, the other envisages sequences of percepts. This second axiom states that, wherever A is found, it will be followed or accompanied by the same thing, B, as it was the first time.

This also depends on the postulate of identity. For if A were followed by B yesterday, and by not-B

to-day, there would have been some relation in which
A stands now, which it did not stand in before; that
is, A would not have been strictly *the same* A in the
two cases. We should find that some *respect* had
been omitted, in which what we now call A was different from what we then called A. But if no such
difference exists, and yet the postulate is true, then
A must be followed by B, both yesterday, and to-day,
and whenever it occurs.

Book III.
Ch. VIII.

§ 6.
General
remarks on
Axioms.

This axiom of the uniformity of the course of
nature is the one more immediately applicable to,
and approximately verifiable by, scientific observation and experiment. It expresses itself in the *circumstances* of a particular concrete thing or event.
An unexpected difference in behaviour of a chemical
substance, for instance, suggests a difference of circumstances under which it acts differently now from
before. We assume that the same thing under the
same circumstances will act in the same way. And
this axiom is verified, so far as axioms can be, when
we have succeeded in finding a difference in the thing
or in the circumstances, corresponding to the difference in behaviour. The axiom of the uniformity of
the course of nature takes nature dynamically, that
of uniformity of nature, simply, takes it statically;
this is the whole difference between the two axioms.

There is another axiom which used to have a
great vogue, and which had the advantage, if it be
an advantage, of vagueness in not distinguishing the
subjective and objective aspects of things,—the axiom
known as the Principle of Sufficient Reason. This
agrees with the axiom of the uniformity of the course
of nature in taking nature dynamically, adding the
notion of it as a series of causes and effects. It is

Book III.
Ch. VIII.
——
§ 6.
General
remarks on
Axioms.

" Everything that exists has a sufficient reason why
it is as it is and not otherwise." Leibniz, who brought
it into vogue, thus enunciates it:[1] "L'autre *principe*
est celui *de la raison déterminante:* c'est que jamais
rien n'arrive, sans qu'il y ait une cause ou du moins
une raison déterminante, c'est-à-dire quelque chose qui
puisse servir à rendre raison *a priori*, pourquoi cela
est existant plutôt que de toute autre façon." It
assumes identity as the logical norm, and requires
for every difference observed a difference of conditions to be discovered.

The axiom of sufficient reason does not distinguish
between cause and reason, objective condition *existendi*
and subjective condition *cognoscendi*. It should have
run, for clearness' sake, 'Everything that exists has a
sufficient condition of its being as it is, and the knowledge of which would be a sufficient reason to us for
its being so.' It was probably in greater part this
vagueness which brought the axiom out of credit,
and caused it to be replaced by the distinctly objective
axiom of the uniformity of the course of nature. But
properly understood there is nothing unsound in the
older axiom.

This difference is clearly brought out by Kant in
his criticism of Eberhard, whom he accuses of ignor-
ing it:[2] "*Every proposition must have a ground*, is
the logical (formal) principle of cognition, which is,
not co-ordinated with, but subordinated to, the Proposition of Contradiction. *Every thing must have its
ground*, is the transcendental (material) principle,

[1] Théodicée, Partie I. § 44. See also his Réflexions sur le livre
de Hobbes, §§ 5. 6. Also his Remarques sur le livre de M. King, § 6.
[2] In his " Ueber eine Entdeckung zur Kritik der R.V. Erster
Abschnitt. A. Werke, ed. Rosenk. u. Sch. Vol. I. p. 409.

NATURE AND HISTORY. 111

BOOK III.
CH. VIII.

§ 6.
General
remarks on
Axioms.

which no human being has ever proved or ever will prove out of the Proposition of Contradiction (or out of mere concepts at all, without relation to sense intuition)." Observe, by the way, that Kant takes the matter, as if to prove the second principle out of the principle, or as he calls it the proposition, *Satz*, of contradiction was the same thing, in point of kind, as proving it out of mere concepts divorced from percepts. This was a consequence of his separation of faculties, of thought from sense. But the postulates of logic, and among them the *Satz* of contradiction, never are divorced from percepts; so that to prove the *ratio sufficiens existendi* from the postulates would not necessarily involve proving it from concepts alone without percepts. This consideration shows how wide is the difference between Kant's way of treating questions of this kind and the way which I call the metaphysical.

The principle of sufficient reason is thus distinguished into two principles, one *existendi*, the other *cognoscendi;* the latter of which is purely logical. And this latter principle it is which is intended by Lange's "*Axiom von der Begreiflichkeit der Welt,*" which he says is identical, in the Kantian philosophy, with the first stage of Teleology. "The 'formal' teleologic character (*Zweckmässigkeit*) of the world is nothing else than its adaptation to our understanding, and this adaptation (*Angemessenheit*) requires just as much the unconditioned rule of the law of Causality, free from mystical encroachments of whatever kind, as on the other side it presupposes the understandableness (*Uebersichtlichkeit*) of things, by their being arranged in determinate forms."[1]

---

[1] Geschichte des Materialismus, Vol. II. p. 276. 2nd edition.

Book III.
Ch. VIII.

§ 6.
General
remarks on
Axioms.

Many scientists are fond of saying that all the truths of science, even the most axiomatic, are contingent. Philosophers are fond of claiming necessity for the laws of thought. Now it is impossible to have necessary laws of thought and contingent things to which they are applied. The contingency of the things would destroy the necessity of the thought; the necessity of the thought would make the things necessary also. You cannot separate, but can only distinguish, the two. They are inseparable. Either both are necessary, or both are contingent.

True, the universe, supposing it (*per impossibile*) to be a finite thing, might become a chaos tomorrow, and then there would neither be laws of thought nor laws of things any more. Both are contingent in this sense. There is no guarantee for percepts continuing to exist, because there is none for existence itself, *on this supposition*. But as between percepts and concepts, both are necessary, both are guaranteed by the percepts which actually constitute the world. The question is, what this guarantee consists in, and how it takes effect.

Here is directly applicable the distinction of nature and history. For from what point of view does the world look to us contingent? From that of its history. There is no guarantee for percepts continuing to exist; that is, there is no *conditio existendi* prior to percepts, prior to existence generally. To enquire into the history of anything is to enquire into its conditions of existence; in the case of finite things, into their *genesis*. In naming the universe, we treat it as finite, though really infinite; and then finding that we cannot assign its conditions of exist-

Book III.
Ch. VIII.

§ 6.
General
remarks on
Axioms.

ence, that is, that it has no genesis, we hold it to be contingent.

On the other hand, from what point of view does the world appear necessary? From that of nature. For the elements which constitute it are infinite and eternal. The infinity and eternity of the *nature* of existence are the guarantee of validity of those laws which are co-extensive with it. In other words, both the laws of thought and the laws of things are guaranteed by the percepts which they modify, and again the same question recurs as to the mode and therefore the value of the guarantee.

There is at any rate one question, relating to the *a priori* or *a posteriori* character of the ultimate postulates and corresponding axioms, which can now be judged. They are not forms in the mind, considered as an immaterial entity, *a priori* to experience. They are bound up with experience itself. They are *a posteriori* to percepts generally, because all percepts are empirical, consisting of inseparable elements; but *a priori* to all portions of conceptual experience, in which they are the foundation or underlying fact.

This same feature, that of being *bound up with* phenomena, meets us again, but with modifications arising from the lower stage on which it occurs, when we turn to the case of axioms less general than those of uniformity. Take for instance, in pure mathematic, the axiom, 'If equals be added to (or subtracted from) equals, the wholes (or remainders) are equal.' This is an axiom concerning quantity, both geometrical and numerical. It is bound up with magnitude. At the same time it is a particular case under a more general identity; it is identity of magnitude, not of quality. In geometry it is approxi-

Book III.
Ch. VIII.

§ 6.
General
remarks on
Axioms.

mately verifiable by actual comparison of percepts in presentation, by superposition.

Equality, then, is a percept which runs through all kinds of object-matter, so far as they have quantity. But now let us come lower down, to geometry alone, where the quantity is particularised as spatial quantity. Geometry rests on a definition and an axiom. But the definition is, strictly speaking, no definition but a *description* ; nor can it be otherwise, seeing that the thing defined is an ultimate *datum*, being one member of a pair of contraries, curved lines and straight.[1] The definition serves merely to *denote* a fact of perception, by describing what we find there. I speak of the definition of a straight line,—*the shortest road from one point to another.* Or, if we propose to define it as the line which lies *evenly* between its extremities, this presupposes the notion of even and uneven, that is, of straight and bent, and is again no definition, but a description as before.

The axiom on which (together with this definition) geometry rests is—*From one point to another, one and only one straight line can be drawn.* The axiom supplements the definition. And it makes no difference whether we call it an axiom or a postulate. In either case, what it does is to show the *reality* of the definition, that is, the possibility of constructing its object in imagination. The definition (or description) of a straight line would be, or at any rate might be taken as, a merely *nominal* definition, without the corresponding axiom or postulate. Coupling them together is the claim made by geometry to be a real science. For in geometry to prove the reality of a

[1] See above, Chapter VI. Vol. I. p. 401.

Book III.
Ch. VIII.

§ 6.
General
remarks on
Axioms.

figure is always to prove that it is possible to construct it, consistently, in imagination. It is indifferent whether we regard geometry as based on a *real definition*, or on a nominal definition and an axiom. In either case the definition is a description of the ultimate perceptual datum, upon which the rest of the science is built.

But again let us come lower down; let us suppose a particular state or grouping of the object-matter, a grouping into resistant solids in motion. This gives us what is commonly known as the material world; particles and masses of resistant matter in motion, in action and reaction, considered as abstracted from their other objective qualities, sound, light, colour, odour, taste. We are in the domain of rational mechanic.

The axioms of rational mechanic presuppose equality in mass, in force, in velocity; and this latter supposes that there are equal spaces and equal times. Newton's three laws of motion, as they are usually called, are called by him *Axiomata sive leges motus*, axioms being the first title he gives them. They are easily seen to be cases not only under the axiom of uniformity of the course of nature, but also under that of sufficient reason. At the same time they presuppose a particular grouping or constitution of the general object-matter; a grouping into resistant solids in motion; the particular condition or conditions of which grouping are not yet, and may possibly never be, discovered by us.

But this want, this failure on our part to assign the particular condition or conditions of the grouping of the general object-matter into resistant solids in motion, or, in other words, of the formation of

Book III.
Ch. VIII.

§ 6.
General
remarks on
Axioms.

"Matter," does not take the axioms of rational mechanic out of the general law of uniformity of the course of nature. It is a defect in our knowledge of the detail, the filling up, of that general axiom.

It is a case precisely parallel to our inability to assign the particular conditions of the arising of consciousness in one kind of moving matter, in nerve substance. The detail of the process is unknown to us; there is a gap between two dissimilars, nerve movement and consciousness, which has not yet been filled up; they face each other "like rocks which have been rent asunder." Solid matter in motion stands as yet like a mountain fastness surrounded by ravines, yet of which we see the affinity of geological formation to the mountain masses which rise on the other side of the clefts. So the mode of formation of solid moving matter out of the general object-matter of consciousness is as yet an enigma; so also is the mode of generation of consciousness from solid moving matter.

The axioms of pure mathematic, of the sciences of calculation and configuration of space, stand on a different basis. They have an object-matter which is indifferent to this particular grouping into solids in motion. Space must be divided by some kind of feeling, distinguished into some kind of figure; but by what kind of feeling, into what kind of figure, is a matter of indifference. Still more indifferent, if possible, is it to calculation, what kind of feeling breaks up continuous time into numerical units; what kind of feeling is the content of the continuous time so broken up. These purely mathematical sciences stand far aloof from, and independent of, solid moving matter and its science, rational me-

BOOK III.
CH. VIII.

§ 6.
General
remarks on
Axioms.

chanic. But rational mechanic cannot stir a step, cannot even exist, without their aid and that of their axioms.

The *scientific* bases of rational mechanic, and of the physical and physiological sciences which are its dependants, must therefore be sought in pure mathematic. Pure mathematic however is a subjective science, in the sense that its bases in turn must be sought in a still more general, subjective, analysis of consciousness. You can have pure mathematic, pure geometry and calculation, without supposing the existence of solid matter; but you cannot have it without supposing the existence of feeling, time, space, the postulates of logic, and the axiom of uniformity.

How narrow and how false, then, is the supposition which the materialistic philosopher makes, that the distinction between solid moving matter and consciousness coincides with that between the objective and subjective aspects of phenomena. The solid moving matter, which contains all the causation which we know of in the universe, is only a portion, and probably a very small portion, of the objective aspect of phenomena. Solid moving matter has a *prius* in existence; it has a *genesis* as well as a *history;* the objective aspect of phenomena has a history only. The scientist is compelled, by his way of distinguishing, to take solid moving matter as *an absolute*, as something which has and can have no explanation; to take it as an object of direct perception, and as an absolute *first* in existence. Thereby he falls into the contradictions set forth by the Kantian Antinomies. Thereby also,—and here is the practical evil, here is the *anarchy* introduced by his

Book III.
Ch. VIII.

§ 6.
General
remarks on
Axioms.

conception,—he has no answer to give to the counter absolutism which he summons into activity, the absolutism which insists that an individual Mind, and not Matter, is the absolute first in existence; the spiritualistic hypothesis, as his the materialistic.

But to return to the axioms. All such axioms as those we have been considering are the first thing, hold the first place, in the demonstrations of the sciences to which they belong; that is, in the sciences considered as demonstrated from principles. Taken *per se*, they are *a priori* to everything else in those sciences. But they do not appear originally *per se*; they appear only as bound up with the percepts, the object-matter of the sciences. In this sense it is that they depend upon, and are derived from *experience*; not in the sense that they are demonstrated or derived from "the particulars," or the facts of the sciences which are demonstrated by their aid. This is why I said in a former Chapter,[1] that Contraries were the same thing as Axioms, but in another shape. The same features in the formal element of percepts give rise to both. Both are expressions of certain inseparable features of percepts, taken in their perceptual character; and both are consequently the foundation of constructive science, that is, not merely of abstract concepts, but of the (Kantian) *construction of concepts*, or positive science.

We reason, then, by the postulates, but without knowing, at the first, that it is the postulates we reason by. We virtually apply the axiom of uniformity and other axioms, ultimate to their own sciences, but not *eo nomine*. Many particular axioms, called afterwards *axiomata media*, are also established

[1] Chapter VI. Vol. I. p. 398.

BOOK III.
CH. VIII.

§ 6.
General
remarks on
Axioms.

as axioms, before we establish the postulates or uniformity as axioms. The *history* of thought is, not that it goes from particulars to generals, but that it goes from the more to the less familiar cases, from Aristotle's γνώριμα ἡμῖν to his ἁπλῶς γνώριμα, though moving always by means of the conceptual order. The order of science, equally moving by means of the conceptual order, is the reverse of the order of history; the order of demonstration reverses the order of discovery. And yet axioms, concepts, and general terms, lie at the root and form the basis of both alike. The first thing framed in order of history of science, *i.e.*, in discovery, is an hypothesis; that is to say, an imagined relation between particular phenomena, based on the axioms appropriate to the case.

We have thus come round pretty nearly to our starting point, the remark with which we began the whole discussion of axioms. The process of discovery is *tentative*, that is, it begins by making hypotheses, and proceeds by establishing them. Both processes are syllogistic, both are inductive. But everything in them is originally tentative, except the postulates and the axioms involved in the processes. But even these, we must remember, have still to make good their claim to unconditioned certainty. With this reserve, the process of acquiring knowledge may be described as a series of guesses, of which those are retained which are verified, and these in their turn become starting points for new guesses or hypotheses.

We have accordingly two *methods* in the acquisition of knowledge, the method of induction and that of deduction. The guesses and verifications which

Book III.
Ch. VIII.

§ 6.
General
remarks on
Axioms.

lead up to a law are the method of induction; those
which are based upon a law are the method of deduction. Both methods make use of syllogism; they
both move forwards by its means, and therefore also
they can be, if needful, tested at every step by being
thrown into a syllogistic form. This and nothing
else constitutes what is properly called the *use* of
syllogisms, that is, their conscious employment as a
means. It consists in their testing the logical validity,
the validity for thought, of the concrete reasonings
and arguments which we frame in following either
of the two methods of acquiring knowledge.

But what more particularly enables syllogisms to
be consciously employed by way of tests? Why are
they not confined to their unconscious use, as in induction,[1] where they simply state the combination of
observations? It is a certain *generality* in syllogism,
whereby it is not confined to facts, nor yet to reasons,
but can combine the two, so as to give at once either
the inferred or the observed reason for an inferred
fact. It is indifferent as to whether its subject-
matter is presentation, or representation, or both.
It can thus combine in its formulas the characteristics of induction and deduction. For a syllogism
is nothing but the simplest way of expressing, the
simplest formula for, a single reason for a single fact
that is not matter of immediate perception. A is D.
Why? Because it is C, and C is D.

Fire is hot, ...   C is D . . . . immediate perception.
Stars are fire, ...   A is C . . . . immediate perception.
Therefore stars are hot . . ∴ A is D . . . . inference not imme-
                                                        diately perceived.

In their formal machinery, syllogisms are the re-

[1] See above, Chapter V. Vol. I. p. 346.

presentatives of the Postulates. They are systems of
propositions expressing judgments, and are knit together by the postulates. They belong to Logic, to
the mechanism of conception itself. And yet they
are not consciously used in the method of acquiring,
but only in that of testing, knowledge.

Book III.
Ch. VIII.

§ 6.
General
remarks on
Axioms.

Taking the view that it is the Postulates which
enable us to frame hypotheses at all, the relation of
Logic to Method, (of which hypothesis is the soul),
is twofold. First, its postulates, by enabling hypotheses to be framed, are the conditions of method
arising; and secondly, its syllogisms are tests of the
correctness of its results. The Methods of science
are built upon the postulates and the axiom of uniformity; their whole framework falls in ruin, if the
postulates are unsound, and we are living in a dream.
It becomes then a matter of extreme interest to determine what is the nature, and what the extent, of
the validity which belongs to them.

CHAPTER IX.

THE POSTULATES AND THE AXIOM OF UNIFORMITY.

§ 1. THE problem which I am now to attack is that which was postponed, it will be remembered, not in the last Chapter only,[1] but in a previous passage[2] where it was represented as one of the chief problems of philosophy, and as soluble only by the aid of the analysis of elements in conjunction with the distinction of the inseparable aspects, objective and subjective. It may be stated as follows: To prove the universal and necessary validity, for all existence, and incapable of any exception, of the so-called axiom, *that the course of nature is uniform.* This axiom was shown in the foregoing Chapter to be the dynamic form of the simpler statical axiom of the *uniformity of nature.* And I choose this form to demonstrate, partly because it is the current form of the axiom, but chiefly because, demonstrating this, I shall have *a fortiori* demonstrated the other, which would not be true *vice versa.* The truth of the dynamic formula implies that of the static.

In mentioning the problem above at the place

[1] See pp. 106 and 108.
[2] Above, Chapter IV. Vol. I. p. 221.

cited, I said, that to prove it satisfactorily, to give a real and not only an apparent and plausible proof, it was necessary to take the question in the whole length and breadth of its difficulty, by stating it in a way equivalent to Kant's way, which was by means of his distinction between synthetic and analytic judgments.

For remark, that the thorough and satisfactory discussion of this, like so many other questions of philosophy, dates back from Kant, and can only be profitably continued by reference to his way of taking it, and the distinctions by which he introduced it; even though we do not adopt those distinctions in our own discussion. The sufficiency of our distinctions can best be made evident by a reference to his.

The axiom of the uniformity of the course of nature, (which for brevity's sake I will speak of simply as the *axiom*), is with Kant an *a priori* synthetical judgment. The postulates of logic, (which I will call simply the *postulates*), are with him the source of analytical judgments only. Now all analytical judgments are with him merely logical and formal, not transcendental and material. They are not constitutive of objective cognition, but a negative or *sine qua non* condition of those constitutive laws. Objective cognition, or knowable existence, is constituted by synthetic *a priori* judgments. We can, then, thoroughly discuss the problem of the inviolable necessity of the axiom only by raising ourselves at least to the height of Kant's position, in his question, "How are synthetic *a priori* judgments possible?" If we would make the axiom depend on the postulates, (which is the course I am going to take), we must at least show that the postulates are equivalent to his synthetic *a priori* judgments, in constituting cognition. Or, if we would prove the necessity of the axiom independently of the postulates, we must show that there are objective axioms, at least as valid as Kant's synthetic *a priori* judgments, and that the axiom is one of them.

I have just said, that the axiom was with Kant a synthetic *a priori* judgment. But this needs an explanatory comment. It was Kant's own theory that raised it to this rank. The Principles (*Grundsätze*) of synthetic judgments, together with their ultimate Principle, *viz.*: "Every object stands under the necessary conditions of the synthetic unity of the manifold of intuition in a possible experience,"[1] are the Kantian analysis of, or rather what Kant substitutes for, the axiom. And in that character they depend upon the Categories, and because of that dependence are synthetic *a priori* judgments. But the axiom taken alone, in its pre-critical state, is no such thing. In an earlier work of Kant's, earlier than the Critic of Pure Reason, but one which contains most of the leading ideas of the Critic,[2] we find the axiom expressed in the words, "omnia in universo fieri secundum ordinem naturæ," and counted as the first of his three *Principia Convenientiæ*, subjective quasi-axioms, the two others being (2) *principia non esse multiplicanda præter summam necessitatem*, and (3) *nihil omnino Materiæ oriri aut interire*. It has here by no means the authority of an inviolable law of existence,

---

[1] Kritik der R.V. System der Grundsätze &c. 2ter Abschnitt. p. 160 et seqq. Hartenstein, 1853.

[2] De Mundi sensibilis atque intelligibilis forma et principiis, § 30. Werke, ed. Ros. u. Sch. Vol. I. p. 340. The date of this Dissertation is 1770.

but solely of a practical precept founded on a large experience. For Kant goes on to say, "Ita autem statuimus [we adopt the axiom] non propterea, quod eventuum mundanorum secundum leges naturæ communes tam amplam possideamus cognitionem, aut supernaturalium nobis pateret vel impossibilitas, vel minima possibilitas hypothetica, sed quia, si ab ordine naturæ discesseris, intellectui nullus plane usus esset, et temeraria citatio supernaturalium est pulvinar intellectus pigri."

Well, but Kant, it will be said, raised it to the rank of a synthetic *a priori* judgment.—Yes, he did so; but observe, in restricting its application to objects of possible experience, and leaving *Things-in-themselves* beyond its range. This distinction is all-important. The transcendental Categories which, according to Kant, are the source of synthetic *a priori* judgments, constitutive of knowable objects, are at the same time the source of belief in *Noumena* which are unknowable objects. While everything in the knowable objects is subject to the law of inviolable uniformity, everything in the unknowable objects is manumitted from that subjection; and the Thing-in-itself becomes thus the possible source of "*causality through freedom*," of which principle Kant made frequent and significant use; for binding *objects* fast in fate, Kant left free the *Ding-an-sich*. The axiom, which before was unlimited in range but not inviolable in operation, became now inviolable in operation but within a limited range. And the problem is, to show how, without restricting its range, without creating Things-in-themselves, an inviolability equal to the Kantian can be secured to it.

We have here another instance of the Kantian

system breaking down over the *Ding-an-sich*, in addition to other cases. which have been remarked. This was the stumbling-block which caused the search after new solutions, Jacobi's, Fichte's. Hegel's. Maimon's, Schelling's. Schopenhauer's, some going back, others forward, but all onwards from the distinction between Phenomena and Noumena as Kant left it.

Quite recently we have had the problem of the inviolable validity of the axiom discussed in this country. and in a way to bring out another side of the question. Mr. Lewes' *Problems of Life and Mind* has been the occasion. The discussion is begun by Professor Bain in the First No. of *Mind*.[1] He there recurs to his own position in his Logic of Deduction, "maintaining that we could give no reason for the future resembling the past, but must simply risk it;" in short taking up the same position as Kant in the passage just quoted from the *De Mundi sensibilis &c.* And he refuses to call the uniformity of nature "an identical truth," as Mr. Lewes calls it; but calls it "a postulate or assumption."

The way in which Professor Bain opens the discussion is this. He quotes Mr. Lewes as maintaining the true expression of nature's uniformity to be: " the assertion of identity under identical conditions ; whatever is. *is* and *will be*, so long as the conditions are *unchanged*; and *this* is not an assumption, but an identical proposition." This is a quotation from Mr. Lewes. Then Professor Bain says: " But now as to the conditions, in what light does Mr. Lewes view *Time* and *Place ?* Are these among the conditions, or are they not? If these are conditions. I fully grant the identity; because the assertion then is that

[1] Mind, No. I. Jan. 1876, p. 116.

what is happening *here* and *now*, *is* happening; and nothing else is happening. But is he prepared to set aside time and place as not being conditions, as not needing to be taken account of at all? * * * It seems to me that to pass the bounds of time and place *is* a hazard; and this is the real point at issue."

Mr. Lewes replies to this point in the next No. of *Mind*.[1] " I answer: Time and Space are abstractions; drawn, indeed, from concrete experiences, but not operative as abstractions among physical agencies." They are *not operative*, that is, not among the conditions *existendi*. Therefore, differences in them, *i.e.*, differences in point of time and in point of space, or place, need not be taken into account at all, in stating the law of uniformity, which thus acquires all the validity of the logical law of identity, notwithstanding that it relates to things existing in different times and places. And the reason given for time and space being inoperative as conditions is, that they are *abstractions*. Mr. Lewes adds, that " the objection, that I do not take into account the possibility that Time may be a condition in causation, has been urged by Professor Clifford, Mr. Pollock, and Professor Bain; and urged by such writers it ought not to be left unanswered." Mr. Lewes' answer, then, is, that he distinctly contemplates and rejects the possibility that Time is a condition in causation, on the ground that both time and space are abstractions; an answer which (if I rightly apprehend it) misses the point of the objection, by understanding it to ascribe efficient, physical, causality to time and space. Whereas it plainly means, that the different times and places, spoken of as possible conditions, may *contain* some-

[1] Mind, No. II. April 1876, p. 283.

thing new which has physical causality; that different times and places may *contain* causes as yet unknown. This objection is not answered by denying that time and space are physical conditions.

The discussion is taken up in the following No. of *Mind* by Mr. Pollock and Mr. Roden Noel.[1] Mr. Pollock argues against Mr. Lewes' position that time and space cannot be among the conditions. He says: "Is it inconceivable, for example, that there should be a secular variation—in other words, a variation depending on time alone as a condition—in the law of gravity? * * * We can only know that there is no sensible variation within the range of human experience as at present ascertained." "So again of space: is it impossible to conceive a minute variation in the law of gravity depending on pure space-relations or on the constitution of space?" And he concludes, "But I fear we are not likely to arrive at any present settlement."

Mr. Roden Noel takes the question up in a thoroughly philosophical manner.[2] Time and Space, he says, "are clearly not *in themselves* conditions that can be taken into account." Why not? Because, as to time, "the conditions to which an ascertained law of nature applies, are *already successive*." Similarly as to space. "The co-existence of phenomena—at all events in the external order—implies Space." To count time and space *in themselves* as conditions would be (if I rightly apprehend the meaning) to count them twice over. They are already conditions of all events and of all objects; therefore Professor Bain's requirement is already satisfied, and conse-

---

[1] Mind, No. III. July 1876, pp. 425 to 42?.

[2] Same place, p. 426.

quently he must admit the validity of the axiom as well for the future as for the present. " It is under condition of difference in time and space that a law of nature is ascertained to be true—therefore, evidently, the difference in time and space cannot suffice to make the same law untrue. Professor Bain says there is no contradiction in supposing that a million years hence the boiling point of water at the ordinary pressure may be raised to 250°. But what *are* a million years *apart from* the uniform succession of more or less similar phenomena, which we mean by a year multiplied a million times? Nothing whatsoever. Therefore, of course, a million years, the other conditions remaining unchanged, can make no difference in any ascertained law or phenomenon." This is a powerful and philosophical support to Mr. Lewes' position. It gives a meaning to his view of time and space being *abstractions*, namely, that they would be so if counted twice over, or as additional conditions.

Mr. Noel proceeds to argue for " rather more than Mr. Lewes allows for" being implied in the axiom of uniformity. I am not sure that I perfectly apprehend his meaning, but it seems to me that what he desiderates is the statement of the axiom as a law of existence, and not merely as a law of logic, a law of what he calls " the essential order of things," an identification of the *ratio sufficiens existendi* with that logical law of identity which Mr. Lewes speaks of. He is afraid that Mr. Lewes' doctrine of the uniformity of nature being an identical proposition will turn out to be only what Kant calls an analytical not a synthetical *a priori* truth, or, if synthetical and *a priori*, yet that it may be found not to extend to the root of things, so as to include things-in-themselves. In short Mr. Roden Noel seems to me to have the very problem before him, of which I am about to attempt the solution.

I have given the account of this discussion[1] at some length, because it opens to us an important side of the whole question. The discussion turns upon whether mere differences in time and space are to be counted among the conditions *existendi* of past or of future objects or events; whether the lapse of time or the change of position in space, *sundered* from any other condition or conditions which might occupy or accompany them, are to be held sufficient possible conditions of a change taking place in the attributes of the phenomenon which is the subject of the lapse or the change.

If such *abstract* Time and Space, as Mr. Lewes calls it, such Time and Space *by themselves*, as Mr. Noel calls it, are, *alone*, conditions of possible change in phenomena,—then creation or origination *ex nihilo* is possible. For this is just what origination *ex nihilo* means, namely, an absolutely *new* thing arising in time or in space. Then, too, Professor Bain is quite right in requiring the express limitation of the postulate of identity to cases where these conditions are *assumed* to be excluded; an assumption, I will add, which will be requisite everywhere and always, not only where an unknown past or an unknown future is in question, but also in the case of every empirically known present, even in the case of the axiom which Professor Bain admits as involving no assumption, "*whatever is—is.*" Then, too, Chance

---

[1] So far, that is, as it had gone, in *Mind*, when this Chapter assumed its present articulation.

is King; and "*causation through freedom*" is no longer a contradiction in terms. But if, on the contrary, creation, chance, freedom apart from law, are strictly speaking inconceivable and unimaginable notions, because consisting of self-contradictions,—then the doctrine from which these notions necessarily follow is reduced, by deducing that consequence, to an absurdity.

But I am not now going to take the apagogic line of proof. I merely point to the necessary logical consequences of holding that mere lapse of time, mere change of place, *sundered* from their content, is, or may be for aught we know, a sufficient *conditio existendi* of change in phenomena. Such a *sundering* of the formal element from the material is an impossible fiction; and its consequences the consequences of a fiction. Creation, chance, freedom apart from law, are changes imagined to be so produced; and consequently are fictions and impossibilities. The question is—Whether the Axiom, or whether the Postulates, override and preclude this possibility, or are themselves overridden by it, and kept to a limited range within it. Are there limits to the Axiom and the Postulates, or are there not? Is there, or is there not, existence beyond their range?

The two sides of the problem now brought out, namely, Kant's distinction of analytic and synthetic *a priori* judgments, and the question whether abstract and separate time and space are conditions *existendi*, afford, when taken together, a sufficient statement of the difficulties attending its solution. If we satisfy both kinds of difficulties, if we look at the problem on both these sides, in solving it, our solution will not have omitted any of the real points which ought to

be taken into consideration. For, by considering it
from the Kantian side, we shall consider whether
anything is left out by the postulates and the axiom,
as being, in point of kind, beyond their range; and
by considering it from the scientist side,—as to whether time and space abstractedly and separately are
conditions *existendi*,—we shall consider whether anything is left out by them as being too minute for
them to grasp. We shall, in short, consider, first,
the *extent* of their sweep; secondly, the *exhaustiveness*
of their grasp within that sweep.

Accordingly, the proof which I am about to offer
falls into two main divisions. Under the first of these
comes the proof that the Postulates of Logic govern
the whole of existence as knowable, that they have a
validity equal to Kant's synthetic *a priori* judgments,
and at the same time a greater extent, inasmuch as
they extend over that which he excluded under the
name of Noumena. Under the second division I
shall show two things, 1st, by analysis of the meaning of the Postulates I shall show, that their grasp
over whatever falls within their range is of an exhaustive minuteness, and that at the same time they
introduce no new element into the perceptual phenomena, except the change into conceptual form ;
and 2nd, that the Postulates carry with them and involve the Axiom of Uniformity, so as to compel us
to regard it as of equal validity with themselves.

So that the proof will practically consist of three
branches, whereby the axiom of uniformity of nature
will be shown to be inviolable and universal, because
it is involved in the postulates which are so. At the
same time it will be made apparent, that the truth
so obtained is solely of a philosophical, not of a scien-

tific character, except so far as this, that it affords a secure *basis* for science; that it in no wise enables us to predict either the occurrence, or the impossibility of occurrence, of any particular conceivable thing; but that in this respect, in respect of the possibility of totally unforeseen changes, which practically and in science is the all-important matter, we remain just in the position which Professor Bain expresses by saying "we must simply risk it." In fact, if the Postulates or Axiom, besides telling us that nature was inviolably uniform, could tell us what phenomena were included, and what excluded, from its actual course, they would be a source of new information; would introduce new cognitions of their own into the perceptual phenomena; in short, would do the very thing on the *not* doing of which their exhaustive validity depends.

§ 2. I. The proof that the postulates of logic govern the whole of existence need not occupy us at great length. It consists in recalling what has already been proved in previous Chapters. It is identical with the proof that the metaphysical method, the method of reflection, is necessary and paramount. In Chapter II. it was shown, that *all* consciousness is included under reflective consciousness; that the notion of "existence" is given only in reflection; and that reflection is a process which presupposes conception, that is, (by Chapters V. and VI.), presupposes the continual use of the postulates. In Chapter III. it was farther shown, that Things-in-themselves were a pure mistake and nonentity, and

Book III.
Ch. IX.

§ 2.
Universal va-
lidity of the
Axiom.

also that whatever was supposed to belong to them is really included in phenomenal existence, the object of reflection. The result of this reasoning is, that, without the postulates, *existence* would be a term *without meaning*. Not only is there no existence beyond the range of the postulates, but the range of the postulates contributes to determine the range of existence, by contributing to determine the meaning of the term. This, then, is my proof of the first point.

II. The proofs under the second division are new, and must be given in detail. And here I must make the preliminary remark, that I shall select the postulate of identity to bear the burden of the proof. It was shown in Chapter VI.,[1] that the three postulates are aspects of each other, expressing *together* one and the same fact, and serving to enforce and explain each other. Now I have to show the identity of the postulates with the axiom of uniformity; and the axiom is a positive statement of general fact. Accordingly the positive postulate, that of identity, not those of contradiction or of excluded middle, is the only one suitable for comparison with it, so as to be regarded as involving it.

I come, then, to the proof of the second point, and I shall begin from Professor Bain's point of view, by showing why it is that, although he expressly admits the postulate as to time present,[2] the axiom appears *not* to be involved in it. This is the case because (as is evident from Professor Bain's language in the passage already cited, "Observation can prove that what has been, *has been*; but it cannot prove

---

[1] Vol. I. pp. 377. 381.
[2] In passage quoted above, p. 126-7.

THE POSTULATES AND THE AXIOM OF UNIFORMITY. 135

BOOK III.
CH. IX.
____
§ 2.
Universal va-
lidity of the
Axiom.

that what has been *will be*") he supposes us to be placed at a point of view looking forward into the future, looking longitudinally not transversely, to go back to a distinction formerly drawn ;[1] and from the shortness of our mental vision unable to *predict* what *will* take place. And this point of view may be taken up equally with regard to objective events (so called) as with regard to subjective perceptions. We appear equally unable to predict what we shall perceive as to predict what will happen. The two things are alike in this. We are then looking at things (or at consciousness) as a *flux ;* and there seems no reason for saying that this flux *must be* uniform.

But this position of looking forward into the future is one which we *may*, not one which we *must*, adopt. The postulates of logic introduce the other point of view, the statical, looking transversely, as it were, at the flux of events and percepts. This is a position which we *must* adopt in reasoning, since we cannot reason without the postulates. We can adopt the other afterwards in addition, if we like, but we cannot do so without having first adopted this one. Perception gives us what we afterwards call a *flux* of objects; the characteristic element in reasoning, which is expressed by the postulates, consists in *arresting* one portion of that flux, making it statical, treating it as a *past*, and then going (not forwards from it) but backwards over it again. The question is no longer, what *will be*, but what *has been*. And this holds good, whatever the duration of the arrested portion may be. A sudden flash, a half-second, an hour, a day, a year, a million years, the whole course of time ;—*everything* is *what it is*.

[1] In Chapter V. Vol. I. p. 370.

BOOK III.
CH. IX.

§ 2.
Universal va-
lidity of the
Axiom.

The next question,—*But what is it?*—is a question no more of divination of a future, but of *analysis* of a past, of a notion which we already have, to some extent, in the mind. We have to break up a static whole into its parts, instead of putting together dynamic parts into a static whole. As, for instance, the year 1870 is the year 1870. But *what is* the year 1870?

I must now recall a distinction drawn at the beginning of Chapter VI.,[1] between the thread of time and the various percepts or concepts in a train of consciousness through which it runs. This time-thread is always going on, whatever may be the content, the states of consciousness, strung upon it, whether percepts or concepts. Now I say that, while the time-thread is always going on, the conceptual content *begins* by an *arrest* of the perceptual content; the moment expressed by the postulates is an arrest of the perceptual content; so that, in the proposition A is A, the second A is the first over again, and not a new percept; and that this is the meaning of the proposition, when taken as a postulate. The postulate of identity does not state that one A is equal to, exactly like, or identical with, another A, but it fixes the first A, whatever its content may be. What we called the flux of the conceptual content is really, in this its first moment, the arrest of the perceptual flux; and it appears to be itself a flux solely because it contains the necessary flux of the time-thread. I proceed to dwell more at length upon this analysis.

Let us go back to the question—*But what is it?*—asked above. This analytical question is proposed

[1] Vol. I. p. 375.

BOOK III.
CH. IX.

§ 2.
Universal va-
lidity of the
Axiom.

to us by the very fact, that in a judgment we have a movement, an explication into three moments of what in perception is but one moment. In A is A, there is but *one* A in question. The form of judgment, necessarily taking place in the flux of the time-thread just spoken of, makes it appear as *two*, subject and predicate, connected by a third moment, the copula. In A is A, we get down to a *single* characteristic, and assert that *this* is *this*, even though it should never recur again. Its only *necessary* recurrence is as the predicate of the proposition, and there it is a *representation* of itself in its occurrence as subject of the proposition. There is but one A. Tap the table,—the tap is A. That tap is that tap. In the predicate, *that tap* is a representation, a representation of *that tap* in the subject.

This being so, the postulate of identity does *not* assert, that the representation is an exact likeness of the original; still less does it assert that exactly similar taps will occur again in presentation. Similar taps, whether in presentation or in representation, would be only partially identical, in quality or kind indiscernible from the original tap. But what the postulate asserts is, that the original tap is itself and nothing else; that by A in reasoning we *mean* A and nothing else. The predicate is a *representation* of the subject, owing to the flux of the time-thread; the *predication* asserts that it *means* the subject (and nothing else), owing to the arrest of the perceptual flux by attention. And in this use and meaning, the postulate of identity, by its extreme generality of form, A expressing *any single* characteristic, enables us to discriminate *in what respects* each similar succeeding tap, when it occurs, departs from perfect

BOOK III.
CH. IX.

§ 2.
Universal va-
lidity of the
Axiom.

identity with the first; but it neither dispenses with actual observation, nor performs it in place of perception. No new element is introduced by it into the percept, but solely a change of form.

The point now insisted on is the real key of the whole solution. The moment of *attention* is a moment of *arrest*. The postulate of identity expresses that moment. The predicate in the postulate is *one* with the subject, because the postulate (though a proposition) expresses the single moment of attention. We are however forced to throw out this single moment into a proposition, because time, the time-thread as I have called it, is essential to all consciousness, to reasoning as well as to perceiving. We can state a *fact* only in the form of a proposition, *one* moment in the form of *two*. But this does not *make* that one moment two; it is one moment in perception, and two when stated in a proposition. Those who think that A is A, the postulate of identity, means that one A is exactly like the next A, whether presented or represented, must, I think, have overlooked the fact of *attention* involved in predication; they must imagine that objects, A, B, C, &c., are given to us *ready known;* and that we have nothing to do but to *compare* them with one another. For if attention is involved in *all* predication, what other function can it have but that of arresting a striking feature in perception; and what other proposition can express this fact of arrest, without mixing it up with other facts, but the postulate of identity?

In another respect also the point now insisted on is the key of the solution. It explains how it is, that the postulate of identity, though it carries the axiom (as will be shown), and though it is valid for past

BOOK III.
CH. IX.

§ 2.
Universal va-
lidity of the
Axiom.

and future equally as for present time, yet does not in the least exclude the possibility of totally new and unforeseen events, but leaves us in the position of " risking it," as Professor Bain says. The reason is, that the postulate, A is A, says nothing whatever of the perceptual content of A. If, on the contrary, its meaning had been, " A now is exactly like A then," we should then be asserting, when we employed it, that our present knowledge of it contains implicitly all our past and future knowledge, all our possible knowledge, of it. It would involve the assumption of an exhaustive knowledge of A, and thus virtually assume *omniscience* in the employer of the postulates. For we should then combine with them the assertion that a particular content was the very thing known as identical; and then the universal applicability of the postulates would turn this exhaustive knowledge of content into omniscience.

Good sense of course revolts against such an assumption. But the escape from being forced to make it, while using the postulates, lies not in restricting the postulates to present or even to present and past time, excluding the future, but in a true analysis of their meaning and assertion of their universal validity in that meaning. Now nothing is said by the postulates as to *what* the content of A is. They express solely the act of *attention*. They are an instrument of infinite delicacy, a knife of keenest edge; and in this sense it is that they are of universal and exhaustive validity, not in the sense of their containing a perception that one content is identical with another. They have no content of their own; they are *employed* as form solely, and not as matter; and the matter which they distinguish consists of the

Book III.
Ch. IX.

§ 2.
Universal va-
lidity of the
Axiom.

percepts strung upon the time-thread in consciousness.

The arrest of the perceptual content, by the repetition of one moment in it,—the psychological condition of which is attention, and the expression of which is the postulate of identity,—is therefore the basis or ultimate *law* of all reasoning. But though it is the ultimate law of all reasoning, it never *appears* in it, and that just because it is the basis *of all*, being involved in the predication made in every judgment. By not *appearing* is meant, that it does not form a link, or premiss, in any chain of argument. But it is everywhere supposed; it is the condition of every term employed in judgment existing as a term. For if we had not fixed on a content as at one with *itself*, we should have nothing definite to compare to anything else, we should have no terms to use, either as subject or as predicate.

At the same time the terms in every concrete judgment are different in some respect from each other; and, of affirmative judgments, only those are at once true and adequate, in which the terms are *indiscernible* except by their position or distance (διάστημα) as subject and predicate. The postulate of identity is in this sense the *limit* of decreasing difference, and of increasing sameness, between subject and predicate. Indiscernibility in point of content is therefore the final test of truth in concrete reasoning, and at the same time rests upon the postulate of identity as its own ground of validity. All that can be done by actual concrete reasoning is to show the indiscernibility of two or more terms for our faculties. A practical and scientific identity is made out, when we can show that strictly no difference but difference

Book III.
Cn. IX.

§ 2.
Universal va-
lidity of the
Axiom.

in time, *numero*, of occurrence to consciousness, as in A is A, exists between the two objects. Imagine this difference also abolished, and you reduce the objects in question to *one*, but then you can no longer express it by a proposition.

The question now occurs, How is the reduction of two terms to indiscernibility accomplished, or, what is involved in the process? In concrete propositions we treat the terms as two, or as different, and try to reduce them to indiscernibility, for that is the proof of the truth of the proposition. What is the method by which this reduction takes place? It clearly consists in analysis. For two differing terms must differ either in respect of their constituent parts, or in respect of their external relations; and there is no indiscernibility of two terms so long as any difference of either kind exists between them. Till then, the object analysed as a whole, the object of the whole proposition, subject and predicate together, is not reduced to unity. For instance, when we say, The year 1870 is the year 1870. Its analysis consists in taking the subject and predicate as two percepts, my notion of 1870 when I say it in the subject and my notion of 1870 when I say it in the predicate, comparing them together and reducing them to one, to indiscernibility, and consequently to identity. This is not done, so long as any shade of difference enters into the one and not into the other.

Whatever may be the object in question, there will always be found, corresponding to every differ-ence of the two notions of it, which we will call AS and AP (A subject and A predicate), there will always be found some difference between the *ana-lyses* of the two notions. For if there were no dif-

BOOK III.
CH. IX.
———
§ 2.
Universal va-
lidity of the
Axiom.

ference in the analysis of their constituent parts, or conditions *essendi*, or as we may also call them *internal* conditions, the two notions would be indiscernible in point of kind; and if, besides, there were no difference in the analysis of their *external* conditions, in the relation of them or of their parts to phenomena outside them, then they would be the same phenomenon in point of number as well as of kind; the year 1870 would be not only precisely similar to itself, but placed only between the years 1869 and 1871. The two notions of the year 1870 would be one and the same notion, the AS and the AP of the concrete phenomenon would be only one A, and the *proposition* "1870 is 1870" would be reduced to the *term* "1870."

The postulate of identity, then, *forces* us to find a difference of conditions, external or internal, for every difference of kind, or general shade of difference, applying to the phenomenon as a whole, in its different times of envisagement. That is to say, in other words, it forces us to adopt the principle known as *Ratio Sufficiens Existendi*, for all analysis.

III. And here I enter upon the third branch of my proof. The *Ratio sufficiens existendi* is the same thing as the axiom of the uniformity of the course of nature. This I now proceed to show.

The axiom has been variously formulated. I will take Fechner's formulation, as one of the clearest, and at the same time not liable to be objected to as unscientific: "Whenever and wherever the same circumstances return (whatever these circumstances may be), then return also the same results; but with other circumstances, other results."[1] Now what is

———
[1] *Zendavesta*, Vol. I. p. 343.

Book III.
Ch. IX.

§ 2.
Universal va-
lidity of the
Axiom.

meant by "the same circumstances" returning? Clearly the same circumstances in point of *kind;* for if in point of *number* of times, they could not *return,* they would not be capable of repetition. But every added sameness in point of kind, whereby one whole set of objects becomes more nearly the same with another set, is an added sameness in point of position of some of its parts relatively to others, or of parts which come into position or which vanish from position relatively to the rest. And for complete sameness in point of kind, there is complete sameness in number and relative position of all the parts. The internal conditions are then completely the same. The same circumstances in the same relative order are repeated.

But this still leaves *isolated* the set of circumstances in question. Their *external* conditions, or their position with regard to other sets of circumstances, are still supposed alterable, without alteration of the set itself. In other words, the set is the same in point of kind, but not in point of number, with itself. If its position with regard to other sets is one of the conditions of its being what it is, then it can only be the same with itself on condition of its holding the same position with regard to other sets, that is, of its being one in number, as well as the same in kind, with itself. In other words, the condition of "the same circumstances returning" is only *realised* by oneness in number, as well as sameness in kind, of those circumstances; and the "returning" can mean in this case only the *mental picture* of the circumstances themselves, the representation of them as being what they are, just as the AP of the postulate of identity is the mental picture of the AS in that postulate. There is but one A involved in

Book III.
Ch. IX.

§ 2.
Universal va-
lidity of the
Axiom.

A is A; the appearance of two arising from the time-thread being necessary to the existence of consciousness, in judgments as well as in perceptions, even in the judgment that A is one and not two. The axiom is therefore strictly true in the same sense as the postulate is, namely, at the limit and as the basis of partial identities. And it follows from the postulate in the sense, that the postulate, when supplied with a perceptual content, becomes the axiom, or, in other words, the axiom is the phenomenal aspect of the postulate.

It no doubt seems, at first sight, a *far cry* from the postulate of identity to the axiom of uniformity. The content of the two seems entirely different. There seems to be a gulf between them, the gulf which Kant discerned between his analytic and synthetic judgments, or the gulf which we all imagine between empty thoughts and solid things, or again the gulf which we imagine between ideal hypotheses of agreement and the endless variety of experience. But the enormous apparent difference arises from this, that we habitually assign them to different spheres, giving the axiom to science, the postulates to logic or philosophy. Now the scientist in stating the axiom *begins* with the tacit supposition that Nature is different, various, changeful, never the same, in continual flux; I mean that he takes it as if it were so, in order to see whether any order and regularity, and how much, can be discovered in it. Whereas the logician, starting from the postulates, and finding that in no other way but this it is possible to think at all, *begins* with the postulates; but the postulate of identity is a statement of complete oneness, sameness *genere et numero*.

Book III.
Ch. IX.
—
§ 2.
Universal va-
lidity of the
Axiom.

Every sameness, therefore, discovered by the scientist, is a case under, that is, an incomplete instance of, the complete sameness *genere et numero* demanded by the postulate. Similarity is partially realised sameness; sameness is partially realised oneness. And just as there are metaphysical elements, always distinguishable never separable, in percepts as such; I mean the elements formal and material; so also in concepts, as such, there are metaphysical elements, always distinguishable never separable, namely, the determinations subject and predicate, which in the postulate of identity we have called AS and AP. Some subject element and some predicate element are requisite to every concept, and the judgment is their combination.

It has been already remarked that magnitude is indifferent in logic; whether we take a large or a small portion of the perceptual train, the method is the same. The logician, then, starting with the postulates, starts with the notion of one and the same *universe*, every part of which is *one and the same* with itself. The scientist starts with the notion of changeful, different, phenomena; groups them into sets of similar phenomena; finds repetitions or recurrences of similar sets; analyses these sets up to the point of indiscernibility; retains the notion of recurrence of sets the *same in kind*; and, carrying up the notion of sameness to completion without discarding the notion of recurrence, states his axiom of uniformity as discovered in nature, without noticing that it is never realised except at the limit, or under the supposition, of oneness taking the place of recurrence.

It is the notion of recurrence of phenomena that

Book III.
Ch. IX.

§ 2.
Universal va-
lidity of the
Axiom.

makes us hesitate to identify the axiom with the postulate. There are no sets of phenomena perfectly identical, unless they are identical in *number* as well as in *kind*, that is, unless they are one and not two sets, unless in fact they are incapable of recurrence. The scientist does not need to push matters to this ideal limit, because he wants his axiom of uniformity for practical application, for application to phenomena considered as a flux of events. But for all that, the ideal limit is a logical necessity, being based upon the postulate of identity, by which alone the scientist or any one else can reason. It is, moreover, his logical justification for using the axiom with restriction to the notion of partial recurring identities. For if the axiom were not perfectly true at the limit, the reasoning under it would not be known to approximate nearer and nearer to truth, in proportion as the degree of sameness becomes greater, or (same thing) the degree of difference less, between the groups of phenomena which he assumes as recurring.

The postulates of logic are their own justification; they neither need nor can they receive any other, except so far as analysis can be called one. I mean, it is impossible to show *that* they must be true; all we can do is to show *what* they mean, that they do not add to, or take from, the percepts which they mould. For as to demonstrating them, they are themselves a prior condition of all demonstration. But the axiom of uniformity practically needs a further justification, as is sufficiently clear from the kind of controversy which exists about its validity. It needs it because, in its shape as an axiom, as expressing the phenomenal (not the volitional) aspect of things, it cannot be verified by presentations of

BOOK III.
CH. IX.

§ 2.
Universal va-
lidity of the
Axiom.

sense. Our senses fall short of perceiving differences of feeling, long before infinity, either in the enlarging or diminishing direction, is exhausted. Experience, therefore, in the ordinary sense of the term, can never fully justify the axiom. It receives its justification from the postulate of identity, by being shown to be its phenomenal aspect; a formulation of it like Fechner's, for practical purposes, is justified as expressing an approximation to the same ideal limit; and that form of the axiom known as the *Ratio sufficiens existendi* is the link by which the connection between postulate and axiom is made evident.

§ 3. I have now given the promised proof with as much brevity as seemed consistent with clearness in a question hitherto so obscure. But there is much to be added by way of elucidation and comment.

In the first place I would remark, that the foregoing theory differs from Hegel, inasmuch as it represents the postulates of logic as superinduced upon and presupposing the flux of perceptions; whereas with Hegel they, or rather that which corresponds to them in his system, produce it; I mean that the force of Negativity, inherent in The Concept, throws up, first, the series of opposites and their combinations belonging to the Logic; then, the same force of negativity throws out the world of percepts; and then the same force combines this same world of percepts again with the world of concepts in the absolute Mind.

Next it occurs to me that an objection may possibly be made, to the effect that logic founded on the

Book III.
Ch. IX.
———
§ 3.
Some objec-
tions con-
sidered.

postulates compares unfavourably with the mathematical Calculus, inasmuch as the calculus applies to
unextended points, or divisions: whereas the postulates seem to apply only to things contained between
points of division, and thus leave an infinitesimal
residuum, an area small indeed but still an area, un-
swept. But the answer is, that the postulates apply
to both. Their symbol A may be taken to mean a
point of division, as well as the space, or duration,
between two divisions. The point of division is in
that case denoted and fixed for thought by naming
the two things which it divides. Thus the postulates
do not assume either that space is not divisible *ad
infinitum*, or that time ever stands still or has no
duration. But they assume, or rather *postulate*, that
any one portion of the flux (whether in space or in
time), or that any point of division in that flux, is
itself, unum numero et genere. The calculus itself is
but an application of the postulates to the measurement of time and space.

This objection is akin to another which has been
already mentioned in the foregoing Chapter, relating
to the impossibility of verifying the truth of any
assertion of strict and complete sameness in point of
kind between two phenomena, owing to the im-
perfection or coarseness of our instruments and of
our senses themselves. I admit that, beyond certain
limits, we cannot verify by presentative sensation the
sameness which the postulates assert. But this is no
objection against the postulates; it has weight only
against the axiom of uniformity when taken as
founded immediately upon observation and experiment. When so taken, and verification attempted
by the same means, the means fall short and the veri-

BOOK III.
CH. IX.

§ 3.
Some objec-
tions con-
sidered.

fication fails. It is this very circumstance which forces us to have recourse to proving the axiom from the postulates. For the postulates do not depend upon that sort of verification, verification by presentative sensation. They belong to the process of redintegration, and compare their moments together in the character of *representations* not of presentations. Differences which are too minute to be presentatively perceived may yet be imagined, and even expressed symbolically; sameness is not perfect which does not exclude every imaginable difference. These are the sameness and difference contemplated by the postulates. And wherever a difference can be imagined, there a sameness can be imagined with equal minuteness.

The process of reasoning, therefore, by these ideal categories of sameness and difference, A and not-A, is a process more minute than the sense presentation by which it is proposed to test it. This test besides is only applicable to one case at a time, to particulars as such. Where it is applied, and a difference is discovered in two things which before were held to be indistinguishable, what follows? Are the postulates, the general law, overturned? Not at all. They are once more found true; for the character of difference depends upon the postulates just as much as the character of sameness. That any percept is found to be different from another, supposes the postulates to be true. For difference from others supposes identity with self. Percepts cannot be known as differents except by being known as self-identical. The postulates are a law not only for presentative percepts but for representative, for imagination in all its fulness and all its minuteness. Thought is involved in all verification, and the postulates in all thought.

Book III.
Ch. IX.

§ 3.
ome objec-
tions con-
sidered.

Another objection may be made. It may be said that, whatever else the postulates may prove, they do *not* prove the inviolability of the axiom; and therefore are of no service to the scientist. For what the scientist means by the axiom is, that the conditions of a particular phenomenon, water for instance, are the same now and next year, the same here and at the Antipodes. Oxygen and hydrogen in combination are the conditions always and everywhere. And it will be said, that I have not shown, by bringing in the postulates, that they *must be so* always and everywhere, which was what he required; as to this, the scientist still has to "risk it," as I have myself admitted.

To clear up this point, it must be observed, that the scientist aims at determining what the permanent conditions of a phenomenon are (say of water), and what conditions may change without affecting the result. Wishing for practical purposes to distinguish the conditions which cannot be changed without affecting the result, it is these alone which he calls *the conditions* of water. But the surroundings are its conditions also, only that they may change without affecting the result in its character as water.

It is a separation of the special conditions, those which cannot change without affecting the result, from the general conditions, the surroundings, which may change without affecting it,—this separation it is which constitutes *science;* and the discovery that the special conditions are the same for the same results, always and everywhere, is the *a posteriori* proof of the axiom, so far as that proof goes.

It is perfectly true that I have brought no *a priori* proof, no proof of the *necessity,* of the axiom in this partial shape. It cannot be a partial *a posteriori* doctrine and a general *a priori* doctrine at once and in the same shape. To become an *a priori* doctrine it must be thrown into a larger form and mould than its *a posteriori* one. What I have done is to show that it cannot be proved true in its partial shape, and why it cannot; but that in its general shape it can, and why it can. To prove that shape of it true which is employed by scientists would involve the assumption of omniscience, exhaustive knowledge of the conditions of observed phenomena. It is the philosophical shape of it alone which is a necessary truth.

This larger and philosophical shape consists in taking *all* the conditions, the general as well as the special, the surroundings as well as the constituents and their combination, and then showing that you cannot imagine *any* change or difference in the result (*e.g.* the water) without *some* difference or change in the conditions; and conversely, that you cannot imagine any difference in the conditions without some difference in the result. And why not? Because it would be incompatible with the *oneness* of the statical way of looking at things in thought. Instead of starting from complete difference of things, you start from their complete sameness.

My position then is this. Instead of saying, The postulates are, but the axiom is not, necessarily true, I say that they are both necessarily true, but that the axiom does not mean what the scientist supposes, namely, that nothing now unknown to us can interfere with the production of demonstrated results from their demonstrated conditions; *e.g.*, that water *must be* produced from combining oxygen and hydrogen in the ascertained proportions, beyond the possibility

Book III.
Ch. IX.

§ 3.
Some objec-
tions con-
sidered.

of its production being frustrated by some condition
now totally unknown to us. No wonder he rejects
the axiom on this interpretation of it.

So far from the axiom implying that we know all
the conditions, it implies the very reverse; it implies
that we do not know them. Indeed, in its first application, the first effort of reasoning, the axiom (which
is involved in the postulates) contains no knowledge
of conditions at all. It is therefore *a priori* to any
demonstrated fact, and consequently prior to all reasonings founded on facts previously demonstrated.
But it gives us no knowledge of what the conditions
are which form its content. The terms *conditions*
and *conditioned* are *relatives*,—that is the sum and
substance of the axiom, and it is a truth of inviolable
necessity.

Between the postulates on the one hand and the
flux of separable percepts on the other, comes the
axiom of uniformity, the phenomenal aspect of the
postulates. It tells us about the course of phenomena
just as much as the postulates tell us about the course
of thought, and no more. It is for perception to tell
us about the content of both. The postulates super-
vene upon the perceptual order, and change it into
the order of thought, and the axiom is the expression
of this resulting order. There is a flux, and there is
order in the flux. The phrase "uniformity of the
course of nature" expresses this double character;
it takes nature as a *course*, and then affirms its uniformity. And the affirmation that the uniformity of
the course of nature is universal and inviolable is a
strictly necessary truth, without the possibility of
suffering exception. An exception would imply the
possibility that the course of nature as a whole could

BOOK III.
CH. IX.

§ 3.
Some objec-
tions con-
sidered.

be an object of thought without being conceived as a
single thing or fixed state.   But the content of the
course of nature might be quite different from what
we experience; and that without in the least affecting
the validity of the axiom.

True, the function of science is to take nature
piecemeal, as it were, to regard it as a flux, and to
observe what appear to be its recurring samenesses or
rather similarities.   It does not need to enquire at all
about the inviolability of the axiom, which is made
safe enough for practical purposes by experimental
verifications.   But it is one thing to go to work practically as a scientist, another thing to deny that the
axiom has any greater validity than the scientist
needs for his purposes, or can make out from his
experiments.   And it is another again, to leap at the
inviolable necessity of the axiom, in its partial shape
as a truth of science; without having traced it to its
source in the postulates, and in this way shown its
purely formal character, as well as the universality
of its sweep.   The metaphysician is concerned to
enquire into the grounds of its validity, and determine whether the uniformity of the course of nature
is merely a practical precept resting on a large induction, or an axiom of necessary truth, involved alike
in the laws of consciousness and of existence.

§ 4. One of the most important questions which
are usually discussed in connection with the axiom
of uniformity is that of the Miraculous, or the possibility of miracles.   Let us see what light is thrown
upon this question by what has now been said.   It is

clear in the first place, that miracles *eo nomine* are impossible, I mean supposing the usual definition of them as *interventions of supernatural power*. For the inviolable uniformity of the course of nature excludes the notion of the supernatural altogether; all power is natural power; and consequently *supernatural power* is a contradiction in terms. It is a contradiction similar to that involved in freedom as causality, or causality through freedom, for all causality is necessary causality, or law of nature. True, there is a sense of the term *freedom* in which it is not opposed to law or necessity, but is a case of it, the highest case of it in point of moral value, that case of it in which liberty itself is the supreme law. This freedom is the freedom of the noblest part of our human nature to act according to its own laws, in emancipation from the tyranny of ignoble or less noble motives. But this is not the sense of freedom in the case now spoken of, in the phrase *causality through freedom.*

And here let me stop for a moment to point out how completely the view now stated satisfies the demands of those noble souls who persist in the fallacy of causality through freedom in moral matters; for it is nobility of soul that moves them, witness Dante, for instance, in the lines:[1]

> Color che ragionando andaro al fondo,
> S' accorser d' esta innata libertate;
> Però moralità lasciaro al mondo.

Though of course, when any question is long debated, there will be ignoble partisans on both sides, on theirs, as well as on ours, who maintain necessity. Their fallacy consists in confusing causality from a

[1] Purgatorio, XVIII. 67.

THE POSTULATES AND THE AXIOM OF UNIFORMITY.   155

Book III.
Ch. IX.

§ 4.
The Miracu-
lous.

higher region than what by many scientists is called nature with causality through freedom, that is, causality from no source at all but merely origination *ex nihilo*. The causality from a higher region, by which I mean the moral consciousness, or in one word *conscience*, fully satisfies the *motive* which leads to the adoption of the fallacy of causality through freedom.

That it does so is shown by the consequences in respect of responsibility. If there is no causality through freedom, they say, then there is no moral responsibility. But I reply, there *is* moral responsibility if there is causality from a higher region; namely, responsibility to that higher region itself. Again, the notion of responsibility presupposes the notion of *law*. And therefore the notion of causality through freedom, which abolishes the notion of law, is really fatal to the notion of responsibility. Just as we are responsible to a human lawgiver, when we are bound by fear of punishment or hope of reward to obey his laws, so we are morally responsible to the higher region of our minds called conscience, and all which it includes, or of which it is the channel, when we are bound by fear of remorse or hope of self-approval to obey the laws of conscience. Responsibility rests on two bases, law as well as freedom. Neither must be omitted in a true theory. And therefore only that notion of freedom can be admitted which is compatible with law, just as only that notion of law can be admitted which is compatible with freedom.

There is no incompatibility whatever between the free action or choice of a conscious agent and the fact of the necessary connection of all phenomena, in the strictest sense of the term; nor would there be any, even if we were to regard the contingent future as not only necessarily determined by the past but as having *already taken place*, a view which, as we shall presently see, is by no means impossible. The larger notion of necessary connection is not infringed, nor its truth invalidated, by the contingency of some of its parts or of the details comprised in it. The two notions, the larger and the smaller, necessity and contingency, are maintained from different points of view, and therefore do not contradict each other. Necessity, the larger notion, is perceived when we look *transversely* at the stream of events; contingency, the smaller notion, when we look *longitudinally*, into the future, from any given moment.

Now freedom of choice or action is a case of contingency; namely, where the contingent event in question is an action of the percipient himself. He is free to choose or to act, when he contains all the determining conditions, or is the sum of those conditions, of the choice or action. This is the case when he acts wholly on motives which are representations; these are ends or final causes of his choice or act. The causes which determine him are internal, not external; in other words, he is self-determined. That is what his freedom consists in.

His *sense* of freedom is another matter. This consists in his ignorance of the conditions, or of their relative force; his ignorance, in fact, of his own motives. When he looks forward into the future, his own future act appears to him contingent; contingent on his will, that is, on the motives and their relative force, which are now acting within him. This fact, the *sense* of freedom, is a fact of conscious-

THE POSTULATES AND THE AXIOM OF UNIFORMITY.   157

Book III.
Ch. IX.

§ 4.
The Miracu-
lous.

ness in its first intention, as I have already observed.[1] It is a fact to be examined, explained, reasoned from, and by no means to be denied. But it does not carry with it the perception or the proof of an *entity* Will, or an *entity* Self, still less of their emancipation from law; it does not justify the absolutist in arguing that we *are* free, in his sense of those words, from the sense of freedom as a fact of consciousness. By *being* free the absolutist understands having some originating agency which escapes analysis; very much as other absolutists attribute to memory a faculty of judging with intuitive certainty, in order to prove that, when we remember (for instance) having been cold half an hour ago, we really *were* so.[2]

That we *are* free, *i.e.*, the fact of freedom, is the objective aspect, neither more nor less, of our sense of freedom, and of itself includes no more than its subjective aspect includes, although it is necessarily brought (as an objective aspect) into connection with its conditions, in order to its analysis. These conditions are satisfied, whenever it is found that a conscious agent, or any portion of his nature which can be separately considered, is determined by motives which are part of the agent himself, (or of the portion considered), and the relative strength of which is unknown to him. In other words, the real and true freedom of a conscious agent consists in his self-determination; and this is secure, without recourse to any absolutist hypothesis.

Neither does the assertion of necessity in the strictest sense give any countenance to what is com-

---

[1] In the Preface, Vol. I. p. 11.

[2] I refer to Mr. Ward's theory discussed above, Vol. I. pp. 254. 274. And see farther, Chap. X. § 6. in the present Vol.

monly called Fatalism. A determination of the
future by the past, as inevitable as if the future was
past itself, is not the assertion, but on the contrary
the *denial*, of the doctrine that a particular event in
the future, if fated to happen, will happen *irrespective
of any particular conditions determining it.* If I am to
be killed, says the fatalist, it matters not whether I
go into battle or not. On the contrary, it matters
everything. If he is to be killed, he will necessarily
do those actions which determine that result. The
actions which he does or omits are the best *evidence*
of what is *necessarily* in store for him. The doctrine
of Fatalism is founded on a fallacy which, so far from
resting on the fact of absolute necessity, can only be
thoroughly exposed by an analysis of that fact, and
on the supposition of its truth. The fallacy consists
in pitching upon a particular and uncertain event,
assuming it hypothetically to be necessarily ordained
to happen, and then substituting our *general* knowledge of necessity for the special, and unknown, *causa
existendi* of the particular event. An *aspect* of the
whole is treated as the *condition* of a part.

But to return to the main subject of this section.
Miracles are another case of contingency. And
miracles *eo nomine*, we have seen, are impossible.
But this says nothing whatever about the events
which have usually been called miraculous in the
past, or of the possibility of events equally strange in
the future. All we can say is that, if they have
happened or shall happen, they cannot be or have
been interventions of a supernatural power. Some
scientists, (by no means all), who happen to be also
non-credents of events which have been classed as
supernatural, wish to keep to the partial, forward-

looking, scientist point of view, and *yet* to deny not only the possibility of these events, which may well be done from weighing the evidence in each case on its own merits, (since after all every tub must stand on its own bottom), but also to deny the possibility of supernatural intervention altogether. But this is logically impossible. They want to make out that we can have *perfect* sameness in two phenomena which are different in time, or place, or both; that is, perfect sameness without reduction to one phenomenon. They want to assert that the laws of nature which are now known to us are all that ever have existed or will exist; so that, if any event contravenes the laws which we now know, it is for ever impossible. This is assuming that our partial knowledge is equivalent to the whole, to possible knowledge as well as actual.

So far from this being the case, it must be remarked that, from the forward-looking point of view alone, taking the flux of percepts by itself, everything is *new*, a new creation. No moment of the flux is repeated; it is a train of *differents*. From the transverse-looking point of view, on the other hand, everything is *old*, already existing, the future as much so as the past. We may picture the case in several ways, and all will bring out what I mean by saying that the two points of view are essential.

Suppose existence to be represented by a mosaic floor, and the present moment of consciousness in a conscious individual by a fly upon the floor. Suppose, first, the fly moving across the mosaic; then the mosaic in front of him represents the future, that behind him the past. Things are in a flux to him; the future is unknown and apparently non-existent.

Or suppose, next, that it is the mosaic which moves, the fly being fixed. Again, things are in a flux to him, the future unknown before him. And remark, that we have only to reverse the position of the fly, turning his face the other way, to be able to represent the future and the past as reversed also. The mosaic then seems to come out into the front from beneath the fly's feet. What has come out is the past, what is to come out is the future; but he is looking *forward* into the past, and has the future *behind* him.

Lastly suppose that the fly is part of the mosaic, and that the mosaic *grows*, beginning from one side of the floor and spreading like a ripple so as gradually to cover it, carrying the fly with it on the crest of the wave. This last picture represents, I think, the ordinary notion which we form of existence, and of the relation of conscious beings to it. We represent ourselves as part and parcel of the onward movement of existence.

But in all these ways of picturing the matter, we have the flux and its arrest equally essential. The *content* of the mosaic belongs to the flux; the picturing it *as a content*, the picturing of any part of it as a part, or of the whole as a whole, belongs to its arrest. And the mere arrest of the flux cannot tell us what the content will be; cannot tell us, in ordinary phrase, that the future will *resemble* the past. There may be some unsuspected black lines in the spectrum. What the moment of *arrest* does is to show that the whole flux can be pictured as one thing, as the mosaic of the illustration; in short to render my two first ways of picturing the mosaic as legitimate as the last, which is the ordinary way.

I do not of course mean to say that the scientists

THE POSTULATES AND THE AXIOM OF UNIFORMITY.   161

Book III.
Ch. IX.

§ 4.
The Miracu-
lous.

wish to keep to the forward-looking point of view *alone;* but that they do not effect a sufficient fusion or combination of the two. Their combination is only a partial one; they carry it to the extent of *recurring samenesses;* a conception in which, as I have tried to show, there is no abiding, because it involves a contradiction. The combination of the two features in thought, the flux and its arrest, must be brought about in a far more radical manner.

The question as to the actual occurrence of the events which are usually called miracles is no doubt a question for science, and not for philosophy. It resolves itself in every case into a question of *evidence* for and against. And this is the ground taken by most of the abler impugners of the belief in those events, following in the track of Hume. They leave behind them the philosophical question of the possibility of miracles *eo nomine.* Now just upon this question it is that light is thrown by discussions like the present of the axiom of uniformity. Nor is the contribution unimportant which is thus made towards settling the question of the miraculous in both its branches. For if the notion of supernatural power or of causality through freedom is left standing, as if these were realities and not self-contradictions, then a *vera causa* of the events usually called miraculous is left standing; and the existence of a *vera causa* for any event, the occurrence of which is in question, is a circumstance which materially and legitimately influences our judgment in weighing the evidence for or against that occurrence. It is in this way and only in this, that the philosophical part of the question has an influence upon the scientific part, the examination of evidences.

To deny off-hand the possibility of totally unforeseen and strange events, on the mere ground that
they contradict experience, as the phrase goes, is to
attribute to the course of nature, as actually known
by science, the universal uniformity of the course of
nature, as a whole; *pars pro toto*. But let us track
the fallacy to its source. It springs, as appears to
me, from a want of sufficient acquaintance with the
nature of concepts, causing us to use them carelessly
and unguardedly.

It was shown above,[1] that the complete filling up
of the *extension* of a concept has an ideal *limit*, which
is the point of its individualisation into its perceptual
intension. Science, as at any one time it actually
exists, is the extension of the concept, *order or course
of nature*. It falls below the ideal limit of completion; that is, it does not represent the fulness or
whole, but only a part, of the order of nature itself,
which furnishes the intension of the concept, an intension or mass of percepts imperfectly known to us.
That which is at any time known of Nature itself,
the intension of the concept, is science, which is itself
the extension of it.

The fallacy in question consists, (1) in neglecting
the difference between the concept, order of nature
as now known to us, and the same concept at its
ideal limit; and (2) in applying to the order of nature
itself, (the intension of the concept), as now known
to us, what is true of it only at the limit, namely,
the characteristic of perfect uniformity. The concept-name, *order of nature*, contains in itself no indication which of the two senses is intended, and we
easily slide from one to the other, from the sense in

[1] Chapter V. Vol. I. p. 323.

which it means knowledge as it is now, to the sense in which it means knowledge as it is at the ideal limit, and *vice versa*. At its ideal limit, perfect uniformity is truly predicable of the concept, *order of nature;* and thus, having no mark in the term itself, we easily fall into the mistake of predicating this same uniformity of the same concept-name, in its other sense of knowledge below the ideal limit. We mistake a particular instance or state of the general term for the general term itself, and consequently for the true object of that general term. And thus it is that we confuse Nature as we know it now with Nature as it is in truth, or to ideally perfect intelligences.

If we suppose an ideally perfect intelligence contemplating the whole course of nature as it in truth is, at the ideal limit of completion of science, we must also suppose that intelligence to have the whole course, past, present, and future, lying before it *as a present, uno intuitu.* Let us see what means we have, how far we can go, to render the existence of such a mode of consciousness conceivable to ourselves.

When Newton, in the passage already quoted,[1] speaking of absolute, true, mathematical time, abstracted from everything external, and considered only in itself and in its own nature, says that it *flows equably,* he was thinking, I imagine, much more of mathematical than of metaphysical time; he was thinking of it as requisite for mathematic, as an *ens imaginarium* capable of being conceived apart from a content, and yet as an empirical existent, with divisions or rather distinctions of its own, such as mathematic finds it necessary to imagine.

[1] Vol. II. p. 87.

Time in the abstract, or *pure* time, cannot be said to *flow*; because flowing is a *motion*, and motion not only involves changes in empirical states of consciousness, but involves also those changes being in space. Pure time, taken in Newton's abstract way, does not *move*, does not even *change*, but may be said with more propriety at least, though even then not with exactitude, to *endure*, to be *duratio*, as Newton calls it.

Even duration is a predicate which is applied to pure time only by a comparison with space. We draw from that comparison our term *duration*, to express that peculiar property of having a *prius* and *posterius*, a former and a latter, which is immediately perceived in states of consciousness occupying time. Time is an elemental form of consciousness having one dimension only,—length. Space furnishes us with our whole *Logic* of time. Time *in its first intention* is an ultimate percept, not describable by any general term.

Now whence comes change? And first, *what* is it? We have seen above[1] that change in states of consciousness is the first empirical percept. It presupposes, as metaphysical elements of its analysis, both feelings and time. But neither of these is its *conditio existendi*. Both are involved in it as *conditiones essendi*. Change and empirical consciousness are coeval and inseparable.

To enquire, then, into the moment of origin of empirical consciousness in a nervous organism is the same thing as to enquire into that of change in states of consciousness, and (in the case of solid matter) into that of motion in space. Let us begin with the last.

———
[1] Vol. I. p. 251.

Suppose a ball suspended by a string in front of a screen, moving as a pendulum, and describing an arc from A to A'. It passes through all the intermediate positions, *a*, *a'*, *a''*, &c. Let us suppose these to be excessively minute portions of the arc, say *minima sensibilia*. The traversing of each one of these minute portions severally and successively is irrevocable; it happens and is gone for ever. There is *time*. But is not each minute portion of the arc, described on the screen, gone irrevocably also? No. We see the arc and the screen all at once, and it *remains*, while the ball moves over it. There is *space*. The space remains in time; the traversing vanishes in time; but what of the moving body, the ball?

I mean, is it one and the same ball in *a*, *a'*, *a''*, &c., or is it a slightly different ball in each portion, or is it a wholly different one? Physically, no doubt, it is slightly different. Perceptually, it is wholly different, being perceived at different times. Conceptually, it is wholly the same, since logic abstracts from all difference but difference *in kind*. Of the three, *perceptually* is the mode in which it concerns us at present. There is a ball at *a*, a ball at *a'*, a ball at *a''*, &c.

Now let us make the attempt to suppose all these balls, or if you like *states* of one ball, to be drawn out, so to speak, so as to form a coexistent series, just as the portions of the arc, *a*, *a'*, *a''*, &c., coexist. Instead of *vanishing*, let the movements be (if possible) preserved, as a series of balls or states of the one ball. The movements of the ball *in time* would thus become *fixed*, as a series of permanent states in *space*.

Is this a possible supposition? Clearly not. And why not? Because it contravenes the fact of the impenetrability of physical matter. The balls would have to interpenetrate one another, or parts of them would have to interpenetrate other parts, unless they were reduced to mathematical points, which is inadmissible. We cannot therefore imagine the movements of solid bodies preserved in time, as a series of coexistent states, *unless* we can, either by imagining a *new capacity* in space, or by some other means, get rid of the impenetrability of physical matter. Whether any such enabling hypothesis is possible, is a question not for science, not for metaphysic, but for the constructive branch of philosophy.

But it is different when we come to enquire into the two first of the three cases mentioned, empirical consciousness and change. These are coeval with and inseparable from each other. And they are to be considered as arising in, and depending on, the motions or tendencies to motion (of some sort or other) in solid matter, namely, in the nervous organism. These motions and this matter take the place of the moving ball in the foregoing illustration. But the changes and the conscious states, attached to them and depending on them, may now, if we choose, be very well imagined as preserved in time, and forming a series of coexistent states, the *objects* or *objective aspects* of the successive and vanishing states of presentation attached to the successive and vanishing motions of the nervous organism. We are not restricted to imagine these objective states as arising and vanishing with the subjective states corresponding to them, but may imagine both the past and the future existing permanently, waiting the summons of their condition, the causal movements of the nervous organism, in order to be, the future called, the past recalled, into consciousness. Subjectively they would arise and vanish with those movements, but objectively they would exist permanently; the future would have already taken place, just as much as the past.

I said above[1] that I should recur to this point; and the foregoing remarks may serve also to explain my illustration of the mosaic pavement. We are at liberty to consider the phenomena in this way, because we are ignorant of two things, 1st, of the mode of connection between consciousness and its organism; 2nd, of the conditions which determine the origin of solid, physical, matter, of which the organism is a case. From subjective analysis we know that physical matter is a combination of visual and tactual sensations. But we do not know what determines this combination, nor do we know whether a consciousness may not be possible on other conditions than that of a physically material organism.

We are accordingly at liberty to imagine states of consciousness objectively existing without that condition, and prior to the particular combination of them into the states of physical matter. But observe, any hypothesis as to the mode in which this combination takes place, or any hypothesis as to other conditions of a consciousness besides physical matter, would be an hypothesis, not in science, not in metaphysic, but in the constructive branch of philosophy. It would require, at least, a speculation either as to a fourth dimension, a new capacity, in space, or else a new capacity in visual and tactual

---

[1] At p. 156.

sensations, by which they could be combined without the property of impenetrability, so as to *intensify*
the content of space without recourse to the notion
of crowding or pressing, whether it be a crowding of
parts external one to the other, or a crowding of
vibrations increasing in velocity. But be the hypothesis what it will, whatever it may involve, it will
not be metaphysic, because it will not be analysis, and
it will not be science, because the causation will not be
of the only kind known to us by actual experience.

I am not, of course, maintaining that the supposed fixity of the objective aspects of conscious
states in time is the truth; but merely bringing
forward the conception of it as a means of enabling
us to imagine how an ideally perfect intelligence
would perceive his universe. He would perceive it
as an eternal picture, with infinite difference of parts,
but difference without change. We on the contrary
perceive no difference but by means of, or presented
in, change in states of consciousness.

The apparent obstacle, again, to our getting rid
of this necessity of change consists in the fact that
our consciousness is bound up with a physically
material organism, the movements in which must be
vanishing and successive, owing to the impenetrability of the matter.

Three problems are thus suggested to our enquiry.
The first concerns the mode of connection between
consciousness as we know it and the processes of the
living material organism upon which it depends.
This is a problem for psychological science. Both the
terminus a quo and the terminus ad quem are known
phenomena; we have to discover the mode or the
circumstances of their connection.

The second problem arises out of the first, and carries us at once into the constructive branch of philosophy. It relates to the conditions of arising of physical matter and motion; matter and motion being, in all probability, as we have seen above,[1] coeval and inseparable; and the former problem leading up to this, because the nervous organism is a case or instance of physical matter in motion.

The third problem carries us still farther into the constructive branch. It relates to the condition of change itself in consciousness, subjectively taken. Matter in motion is the apparent obstacle in the way of our imagining consciousness to be fixed; but this does not imply that matter in motion is the sole condition of change in consciousness; there may be other conditions, relating to consciousness at large, as well as this condition which relates to consciousness in dependence on a nervous organism. In fact, considered subjectively, change is the larger notion or fact, of which motion is a special instance,—since it is at once change in space as well as in time. Consciousness in its lowest terms, as actually known to us, is a succession of different feelings, as we saw in Chapter IV. If, then, it is possible, as we have seen that it is, under certain hypothetical circumstances, namely, *at infinity*, to sunder change and difference in consciousness, to conceive consciousness of difference without change, then the further question arises, —What are the conditions of difference being perceived as change, of different feelings coming to be perceived as vanishing and successive?

It seems as if this question touched the very core of the difference between a finite imperfect intelli-

[1] Vol. I. p. 266.

gence and an infinite intelligence of ideal perfection. Until *infinity* is reached in knowledge, however large the object, however delicate the sensibility, of presentative perception, change remains an inseparable feature of all empirical consciousness, and representation remains as essential an ingredient of it as presentation.

CHAPTER X.

CERTITUDE AND TRUTH.

§ 1. To ask if there is a Constructive Branch of Philosophy, if there is an Unseen World, beyond this of ours in which all things depend for their existence upon solid moving matter, an unseen world which yet we can and do know something of,—to ask this question is one and the same thing as asking whether Time, Space, Feeling, the Postulates of Logic, and the Axiom of Uniformity, or any of them, have universal and necessary validity, that is to say, a validity beyond the particular combinations or instances of them with which we are acquainted in the actual world.

We may put this same question in yet a third shape: Beyond direct perception, is there a world embraced by reflective, and do we know anything of it?

To Hegel there is no question of this kind possible. Or rather it is answered, in the affirmative, before it can be asked. In other words, the answer is assumed by his method of analysis. His *Absolute* obliterates the distinction between direct and reflective perception, upon which the question rests.

Hegel's universe may be pictured, though solely by way of illustration and to aid our comprehension of it, as follows.

Out of space, out of time, there is a point of logical contradiction, dividing an infinite line into contradictory parts (representing terms in a proposition). As opposites these parts are incompatible terms; but united they are one term. One is *existence*, the other *nothing*. The point of contradiction both divides and unites them. Passing from one to the other across that point, they *repel* each other; but still they are parts of *one* straight line. Imagine, then, that *united* they form a new line, on one side only of the point of contradiction; so that this new line or term, which is Hegel's *Becoming* (*Werden*), or rather its result, *Something* (*Etwas*), is like a radius, while the two first together are a diameter. The negativity makes its diametrical terms into radii, and each new radius into a new diameter; in short becomes the self-turning axis of a wheel, producing as it turns, and filling an infinite circumference with pairs of contradictory and infinite radii. These are the Concepts of the Logic.

When the wheel is full of spokes, that is, at the *limit* of completion of the Concept, imagine it to revolve on the last and longest of its diameters, at right angles to its first axis; it will then describe a sphere, filled with infinite *planes*, in infinite number. These are the concrete existents of the Philosophy of Nature.

But again, at the *limit* of completion, the sphere revolves, on an axis at right angles to the last, and fills itself again with a new content, compounded of the two former, the abstract concepts of the logic together with the perceptual existents of nature. It now represents an infinity of actual self-conscious beings, comprised in an absolute being, the object of the Philosophy of Mind. The centre of the whole is the principle of contradiction, the eternal well-spring of thought, motion, and life; and reflection is bound up, and assumed along with the principle of contradiction, just as every other kind of content is.

The possibility of a world of reflection, as distinguished from one of direct perception, is not a *question* in a theory of this nature; that world is already actual; "contradiction" or negativity, in assuming a perceptual content, obliterates the distinction between the two kinds of perception, and the function of reflection would be represented, in our imaginary sphere, by the welding into one, or identifying as one straight line, any two opposite, or contradictory, radii. The Hegelian school, then, answers the question without asking it, by an assumption which precludes its arising. There is another school, that of Comtian Positivism, which neglects the question as otiose and mischievous. There is another school, that of Hamilton, Mansel, and Mr. Spencer, which holds that the question is answered by an Unknowable Existence. There is a fourth school, which I have already called English Positivism, which holds that there is no such question at all; science is to it philosophy. There is yet another school,—though not primarily a school but a Church,—which makes its profit out of these divergent opinions and the lacunæ which they leave in knowledge; which profits by them to establish a foregone conclusion of its own, the conclusion which is requisite for its existence as a Church, the conclusion of a definite and supernatural *revelation* from the unseen world to the seen.

A church philosophy is no philosophy, but the antagonist and denial of all. It is so by an ineradicable necessity of its nature; it can cease to be anti-philosophical only by ceasing to be what it is, a church philosophy. And for this reason, that its existence as a church philosophy involves the one sole feature which is the contrary of the search for Truth, the feature of making a particular doctrine a *sine qua non*, the feature of maintaining a foregone conclusion. In vain do the philosophers of such a pretended philosophy appeal to the apparent obviousness or necessity of the principles upon which they build; their foregone conclusion vitiates their reasoning *ab initio*. They love what they are pleased to call Christianity better than truth; they are Christians (so called) first, truth-seekers afterwards, they love truth *incidentally*, κατὰ συμβεβηκὸς, because what they call Christianity happens also (in their opinion) to be true.

I am not alluding in these words to the openly avowed submission of the results of reasoning to the dicta of an infallible church, so frequently to be met with in the writings of Roman Catholics. It is a far deeper vice, and having therefore a far wider influence, than this, which is but one result of it, that I have now in view. When once an infallible church has been admitted to exist, then the members of it *must* suspend their own judgment, the judgment of any individual, upon its infallible dicta, in any matter where it pleases to pronounce. This is merely the *fruit* of the principle which I speak of. The question is, how comes an infallible church to be admitted at all? The logical requisite of it is the belief, that a distinct and particular revelation of events and doctrines has been and is being made from the unseen world to the seen.

The *fact* of such a revelation it is sought to prove partly by occurrences said to have taken place in the seen world; and partly by *a priori* reasoning, reasoning of a pseudo-philosophical kind (there being a foregone conclusion to establish), directed to show first the possibility, and then the probability of the revelation.

This is not the place to insist upon the reasons which have led the spiritual nature of man to revolt against the *conclusions* of the existence of an infallible church, or of any church existing as the depositary and mouthpiece of such a revelation as I have described. The practical consequences which have resulted from a body of men arrogating the function, and armed with a corresponding power, of maintaining a doctrine supposed to be absolutely true and necessary, have over and over again called out, thanks be to God, the most determined spirit of opposition. The whole spiritual nature of man revolts against the tyranny. As a social, as a moral, as a religious being, no less than as an intellectual one, the whole being of man rises in the holiest of rebellions.

But what I have here to do with are the logical principles upon which this conclusion rests. Nor yet with all of these, but with those of them only which are, or claim to be, of a philosophical character; that is, not with the historical occurrences which are made part of the proof of the supposed fact of a revelation, but with the grounds of its possibility and probability which must be sought for on philosophical territory.

Philosophy has the duty of repelling these incursions, of keeping itself untainted by the vice of foregone conclusions, of refusing the sanction of its sacred name to principles and methods adopted for any other purpose than that of searching for truth.

What then are the principles and methods of this antiphilosophical philosophy, by means of which is established the possibility of a distinct and particular Revelation from the unseen world to the seen? I use the words *distinct and particular revelation*, rather to designate and denote than to define or connote what I mean. In one sense, all feeling, all consciousness, is revelation; objects are revealed by sight and touch; the sun in heaven and God in the unseen world are revealed by their own light. But it is a different kind of revelation that is now intended, it is a revelation of particular *doctrines* supposed to have been made to a particular set of men, and continued to their successors; a revelation of the decrees of God and of his dealings with the world, beginning with his creating it. Henceforth in this Chapter when I use the term revelation alone, it will be in this latter sense.

The problem is—in the whole intricate labyrinth of the existing scholastic philosophy (so called), to put our finger upon that or those fontal fallacies which are the basis of the system. The system itself consists of a series of explanations, often of extreme subtilty, directed to get rid of difficulties, which arise in the first instance from the fontal error, and in the following instances from each preceding explanation; each explanation introducing a new difficulty requiring a new explanation. The entire series of explanations, introduced to elude difficulties which ought never to have arisen, simulates the character of a system of philosophy; in reality it has proved nothing, but (at best) has removed objections to its own first principle or fontal error. It is a system of hypotheses devised to make apparently tenable the single principle or principles, which thus are its conclusion as well as its basis. What is this fontal error?

It is nothing else (as I shall proceed to show) than treating Existence as an object of direct instead of reflective perception. Existence is by this means made an *absolute*, taken as something known because familiar, as something of which we know *that* it exists without having any knowledge of *what* it is. In one word it is the assertion of existence as a *Thing-in-itself*. To know anything about *what* such existents are, since we have by hypothesis cut them off from ordinary knowledge, requires a supernatural intervention, in other words a Revelation. Thus the basis of Church philosophy and scientist philosophy is the same. Both proceed from the same fontal error, of putting direct consciousness in the place of reflective. And both are equally opposed to the Philosophy of Reflection.

The whole of ordinary knowledge bears this revelationary stamp, bears the character of a discovery of something absolute, separate from the mind that discovers it. The mind itself is known in the same fashion. *Here* stands the mind, *there* other existents. It also is an absolute, and is known to itself as such; we are supposed to know that we *have* a mind, before we know what that mind is like. The mind is revealed to itself, and other existents to the mind. In ordinary knowledge there is held to be a certain machinery by which this revelation is effected; the mind works in certain ways, works by means of a certain set of most abstract concepts, by which it "cognises" the "essence," the "real ground," the "substantia," of itself and of other existents. The correspondence between these concepts and the "essence" of the existents,—how comes it about, what guarantees it? Nothing. This is what I mean by calling it a *revelation*. Both the *that* and the *what* of existents, as well mental as bodily, are revealed in this sense. For these existents are first *assumed* as existing *per se*, (revelation the first); and next they are cognised by means of concepts *assumed* to correspond to their real and inmost nature, (revelation the second).

This being supposed to be the character of ordinary knowledge, it is easy and indeed necessary to apply the same rule to supra-ordinary, to the knowledge of the unseen world. In ordinary knowledge we have a content of sense perceptions to begin with, to which our knowledge of real existents by means of concepts is applicable. In supra-ordinary knowledge, we have the knowledge of real existents by means of concepts to begin with, the content of that knowledge must be supplied from elsewhere. It is supplied by Revelation, in the usual sense of the term; a revelation from the Beings of the unseen world, a revelation not to the bodily but to the mental eye.

The whole weight of this theory rests, not on the possibility of this latter revelation from the unseen to the seen world, but on the possibility of the two former revelations (so called by me); in other words, on the truth of the theory of cognition which it involves. The question is—Do we cognise, by most abstract concepts, the "essence," the "real ground," of existents? Or is this essence, this real ground, a pure fiction, resulting from an hypothesis, from an hypothetical mode of cognition, adopted to explain the fact of knowledge on the assumption of a separation between the thing knowing and the thing known? The gist of the whole matter is in the analysis of the phenomenon of cognition.

Once assume the separateness, as well as the distinctness, of things knowing and things known; once assume that existents are absolute, things-in-themselves, objects of direct and not of reflective perception; and almost any hypothesis to account for their combination in cognition will become plausible. They, or the words which stand for them, are so clearly combined in cognition, that the mind is tempted to regard as superfluous any theory as to how they are combined; we are impatient of explanations of a fact, which as a fact is so obvious. Idle futilities, word-making and word-fighting, we exclaim in disgust. Meantime the Scholastic goes on his way rejoicing, with his assumption as good as granted, his theory as good as received, the scholastic philosophy as good as established; at any rate enabling him to make the claim, that the doctrines of the Church have the support of reason and of philosophy.

§ 2. The only church that has a philosophy distinct from, but yet wholly at the service of, its theology, is I believe the Church of Rome. Its philosophy is the scholasticism of St. Thomas Aquinas; which within that church flourishes in full vigour, although not much regarded by the rest of the world, which from the era of the Renaissance downwards has preferred to philosophise without the trammels of a foregone conclusion. Perhaps no better authority on the nature and scope of this scholasticism could be found than the able and learned work of Father Kleutgen, "The Philosophy of the Foretime Defended;"[1] a work which, although primarily directed against dissidents from this pure philosophy within the fold of the Church, yet combats their arguments by exhibiting in its completeness the whole edifice of scholasticism from base to roof.

The "Philosophy of the Foretime" has three main divisions.[2] The first treats of the Intellectual Perceptions (*Vorstellungen*), and of Realism, Nominalism, and Formalism, in other words, of the Theory of Cognition. The second treats of the basis of philosophy, that is, of Certitude; of the Principles of Thought; and of Method. The third treats of the content, or truths established by philosophy, including them under the four heads, Existence, Nature, Man, and God. The doctrine of Certitude, which is here treated in the second place, is that part of the whole system which prepares the ground for the doctrine of Revelation. Hence it often holds the foremost place with Scholastic writers; for instance in Father Balmes' Philosophie Fondamentale;[3] and

[1] Die Philosophie der Vorzeit vertheidigt von Joseph Kleutgen, Priester der Gesellschaft Jesu. Zugabe zur "Theologie der Vorzeit." 2 vols. 8vo. Münster 1860 1863. With several Appendices (*Beilagen*).

[2] Work cited, Introduction, Vol. I. p. 22-3. With the Table of Contents.

[3] Philosophie Fondamentale par Jacques Balmes. Traduite de l'Espagnol par Ed. Manec. Paris, 1852.

the Introduction to Mr. Ward's Treatise on Nature and Grace.[1] And it forms the entire content of Dr. Newman's acutely argued work, The Grammar of Assent, certitude being a complex assent.[2]

But the true place of the doctrine of Certitude in a complete system of scholasticism is, I think, where Father Kleutgen sets it, that is, in dependence on the general theory of cognition. For we find that, when the *truth* of our certitude comes into question, this has its basis only in the doctrine of self-cognition, of that cognition of our own "essence" (*Wesen*), whereby we are supposed to know, not only that we know anything, but also that we know the truth in that knowledge. In other words, the doctrine of Certitude depends on the kind of knowledge we have of our own being.

Thus we read in Father Kleutgen: " But we find in the Foretime also the express doctrine, that the mind is made capable of Certitude solely by this, by its cognising not merely its activity but also its essence (*Wesen*). Indeed it is just this which St. Thomas means specially to prove in the passage above cited. We become certain by this, namely, by cognising the truth of our cognising, that is, the agreement of our thought with the existence (*Sein*) of things. But this the mind could not do, if it cognised merely its activity, and not also the nature of that activity ; again it can cognise this nature of its activity only because it cognises the nature of the active principle, *i.e.*, itself ; having the insight (*einse-*

[1] On Nature and Grace., By William George Ward, D.Ph. Book I. London, 1860.

[2] An Essay in aid of a Grammar of Assent. By John Henry Newman, D.D. 2nd edit. 1870. See Part II. Ch. 7. p. 203.

hend) that it lies in its essence (*Wesen*) to cognise things as they are."[1]

The whole question of the basis of certitude depends, then, upon the nature of self-consciousness and of self-cognition, that is, speaking generally, of the phenomena of Reflection. The doctrine of Reflection is one upon which Scholasticism very highly prides itself. It claims to have insisted on the Cartesian *Cogito* long before Descartes. Thus we read: "It is made a merit of Descartes' to have recognised that self-consciousness comes to pass, not by a logical inference, but by a simple mental insight (*Geistesblick*). Our readers however will not fail to see, that this is the very thing enunciated in the doctrine of St. Thomas just expounded."[2] And again: "This his doctrine" [that of St. Thomas] "is however in the fullest accordance with the whole cognition theory of the Scholastics, and that view of the *return* of the mind upon itself was so general, that they were wont to designate the mind, in contradistinction to sense, as principium, quod super se ipsum reflectitur, or more briefly: principium reflexivum."[3]

Now it is in this very theory of Reflection, the scholastic theory of it, that I shall attempt to make good my objection made above, that the scholastics treat existence as an object of direct instead of reflective perception. They do so by *assuming* the mind as a substance *before* it is known in reflection; whereby it becomes of small importance what may afterwards

---

[1] Phil. d. Vorzeit, § 107. Vol. I. p. 186. See also another passage to the same effect, § 297. Vol. I. p. 490. "*Wir haben oben vernommen——übereinzustimmen.*"

[2] Work cited. § 105. Vol. I. p. 182.

[3] Work cited, § 107. Vol. I. p. 187.

be held concerning the reflexive functions of the substance so assumed. Unless indeed it were to be found, that reflection furnished *proof* that the original assumption was true, and that the mind was such a substance as they allege. But this is a proof which never comes. Their mental substance, or principle (*Prinzip*) as it is very frequently called, is an assumption from beginning to end, an assumption bound up with their analysis of the reflective act, and then supposed to be proved by that analysis. It is a comparatively easy task to combat scientists, who simply omit the phenomena of reflection from their purview; it is more difficult to disentangle truth from error in the views of such a writer as Mr. Lewes, with his partial insight into the nature of those phenomena. But it becomes a task of no common difficulty and delicacy, to discriminate and discard the assumption of absolute existents in a system which, being professedly based on reflection, must necessarily introduce the assumption in a manner to elude even the acute vision of its own expositors.

Father Kleutgen begins his exposition of this branch of the subject as follows: "The Scholastics teach, with one accord, that our soul cognises itself only through its activity, and not at all through intuition (*Anschauung*) of its essence."[1] It is then no intuition, intellectual intuition of the nature of the mind, that is maintained by scholasticism. Father Kleutgen puts aside this view at the outset.

He then proceeds to distinguish two things, self-consciousness, or actual (*wirklich*) knowledge of conscious states, " die blosse Erfassung unsers concreten und individuellen Seins," and self-cognition, "Er-

[1] Phil. d. Vorzeit, § 102. Vol. I. p. 176.

kenntniss unserer Natur *d. i.,* der Beschaffenheit dieses Seins;"[1] the *perception* of our existence and the *cognition* of its nature. Self-consciousness gives us the *fact* of our existence, "*dass* wir sind;" we perceive our own "activity;" it is a further question, what is requisite to tell us *what* that existence is, "*was* wir sind."[2] The mere presence of the soul to itself does not suffice to tell us this. "True, we distinguish, in the consciousness of our self which we get by this presence, our existence (*Sein*) from our thought (*Denken*); aye, self-consciousness first arises in and by (*durch*) this distinction. For we cannot apprehend the thought as *our* thought, without opposing our existence to our thought. He that says, *I think*, refers thereby the thought which he perceives to himself as the Subject and Ground of it. But since we at any rate know of this Subject and Ground only by means of the activity which is in and comes from it, therefore its nature (*Beschaffenheit*) also can only be inferred by us from the nature of its activity."[3]

Here, then, I observe we see opened two branches of the enquiry; the first, how we come to think our self as subject and ground of our thoughts, in mere self-consciousness; the second, how we come to know the nature of this subject and ground. Father Kleutgen does not make this observation; he passes at once to the second of these questions, without noticing that there are two, and that he leaves the first unanswered, apparently assuming, as something quite obvious and familiar, that mere self-consciousness includes the notions of Subject and Ground.

[1] Phil. d. Vorzeit, § 103. Vol. I. p. 178.
[2] Work cited, § 104. Vol. I. p. 180.
[3] Same passage continued.

He goes at once to the second question. " If," he says, " we would know its nature" [*i.e.*, of our thinking] " we must first sever it from its content, the thought (*den Gedanken*) from the thing thought, and then strip it of everything which is accidental to it. Then first, when we have apprehended it by means of this abstraction, can we cognise on one side its difference from sense cognition, on the other its dependence upon it, and thereby its own proper nature."[1] The same holds good, he says, of will as well as thought; and these together open the way to a knowledge of the nature of the human being. " For since we have now apprehended it as a Subject in which, and as a Principle out of which, this thinking and willing comes, we infer (*erschliessen*) from this as well its immateriality as its substantial union with the body, and the many other truths which stand in necessary connection with these. For which reason St. Thomas says, that the soul comes to the cognition *that* it exists by its mere presence in its own activity, but comes to the cognition *what* it is only by careful and acute investigation. And he appeals at once to experience. For if the soul cognised by that its presence, not only that, but also what, it is, then this cognition would be just as easy, clear, and certain, as the other. But no one has ever doubted, and no one can doubt, whether he exists, lives, or thinks; while we find out how the living, thinking, and willing principle in us is constituted, only by painstaking enquiry, and in this enquiry are exposed to the danger of error and of doubt."[2]

Now the whole of this reasoning really belongs

[1] Phil. d. Vorzeit, § 104. Vol. I. p. 181.
[2] Same passage continued.

to what I have called the first branch of the enquiry,
—namely, how we come to think or imagine ourself
as subject and ground of our thoughts in mere self-
consciousness. It is this which gives us (if anything
does) the conception of the soul as a substance at all,
—an immaterial substance as we infer afterwards.
Now I maintain, that to know *that we exist* is not
identical with knowing that we are a *substance*. To
know the latter would be to know something of *what*
we are. But in the passages quoted, the knowledge
that we are a *substance* is identified with the knowledge
that we *exist;* and all the undoubted self-evidence of
this existence is attributed in consequence to the
notion of the *substance* which is supposed to exist.
The soul as a substance, as the " subject and ground
of thought," as Father Kleutgen puts it, is not a self-
evident thing. It is the feelings and thoughts, which
are the content of perception in self-consciousness,
that are self-evident; and these it is that give us the
whatness, not of the " principle" of self-consciousness,
but of self-consciousness itself. The scholastics assume that it is possible to know *that* a thing exists
without knowing anything of *what* the thing is. For
they reckon the *substantiality* of the thing to its *thatness* instead of to its *whatness*. And then, since no
conscious being can deny that his consciousness
exists, they claim the admission as evidence of the
substantiality of consciousness, evidence of what
ought properly to be called the *nature* of the thinking
principle.

Nor does it make their case a bit the stronger to
show, as Father Kleutgen does, that they use almost
the very words of Descartes' *Cogito ergo sum*. For
the question is not whether those words are used or

not, since they are of difficult interpretation; the question is, what sense the user attached to them.[1] Descartes used them to recall attention to the phenomena of reflection; to make the interpretation of the *sum* depend on the *cogito*, and not like the scholastics to make the interpretation of the *cogito* depend on the *sum;* to bring the as yet unanalysed *existence* under the grasp of subjective *thought*, not to subordinate subjective thought to the unanalysed and therefore absolute existence; in short, as I have argued in a previous chapter, to equate *esse* with *percipi*, not to equate either *esse* as a whole (as some idealists would), nor yet any mode of it (as the scholastics would), with *percipere*.

But it may possibly be argued, that I am wrong in supposing the Scholastic theory to identify the substance with the *that*, the mere existence, of the soul; but that on the contrary it considers it as one of the first traits *thought* (in thinking the soul), or as the basis of its *whatness*. 'You must admit, they may say, that phenomena have *some* ground of consistency and connection, that actions suppose *some* agent, attributes *some* substrate. This ground, agent, or substrate, (call it what you will), is, they may urge, the first thing in *thought*, the first *grasp* we have of perceived phenomena, and is not due to the phenomena as such. We have, then, in all thought as its basis, they may say, this concept of *substrate* or *ground* or *agent*. Everything must have *some* ground; the universality of the notion makes it quite indifferent whether we count it to the *thatness* or to the *whatness*

---

[1] A discussion (in which I took part) on the true meaning of Descartes' *Cogito* will be found in *Mind*, Nos. IV. p. 568. and V. p. 126.

of any particular thing. The only question is, in the case of any particular thing, what is the next determination of this ground. Some ground it must have; the question is, of what sort. Now, in the case of thought, this ground cannot possibly be material, for thought is very different from sense.[1] It remains therefore that it is an immaterial ground, in other words a mental substance, (*geistige Substanz*).'

My reply to such an argument is the following. Let us suppose, for argument's sake, that the notion of substance is not assumed in perceiving or stating the mere fact of existence, but that it is the first basis laid in thought supervening on that perception. In this case, there is much more in the notion of substance than is either requisite to explain the phenomena, or than can be found in them. As the Scholastics themselves admit, our conception of *bodies* is derived from their accidents,[2] and our conception of souls from their actually perceived activity.[3] But whence comes the notion of *substances* underlying these attributes and activities? It cannot, at least in this place, be alleged that the notion is an innate or connate *a priori* concept *in the soul;* for this would imply that the soul was known to be a substance, and we are now enquiring how this very piece of knowledge *first arises*. Without the prior notion of substance, the dependent notion of *a priori*, innate or connate, concepts is unavailable. This is just where Kant's theory went to pieces. He would have such concepts, and yet would not allow the soul to be a substance. But here it is the position of the argu-

[1] See the passage quoted above at p. 185.
[2] Work cited, § 44. Vol. I. p. 70.
[3] In passage already cited at p. 184.

ment which prohibits our bringing in the notion of substance, simply because it is the notion the origin of which is in question.

Unless, then, we suppose, as I argued above, that the assumption of *substance* was made along with the perception of mere existence, I do not see what sufficient ground there is for bringing in that notion at all. I see indeed abundant opportunity for the mind's inventing the notion, by hasty abstraction from phenomena; but I see no functions actually performed, no phenomena actually taking place, which necessarily demand the notion of substance, and no other, to explain their existence. All the functions of a supposed substance are performed equally well, for bodily attributes, by *space*, which is really found in phenomena; and for thought activities, all are equally well performed by *time*, which is also a reality. Space and time are universally found in these two sorts of phenomena, and the particular space and particular time which they occupy may easily come to be regarded as their substrates, so soon as we pass from perception to thought. They are besides *immaterial* substrates, if this should be thought an advantage.

The elements, time and space, then, not only enable us to imagine a substance if we choose to do so, but, when this capacity of theirs is perceived, they supply its place, they perform the function in philosophy which the notion of substance was invented to perform. Philosophy, when once it has perceived the value of time and space as elements of perception, can dispense altogether with the notion of substance; and this is what the system of philosophy, which I now propose, has for the first time done.

In the next place I must remark, that I do not
see the cogency of the argument for the immateriality
of the ground of thought, derived from the difference of thought from sense. *Material* means, in its
true sense, any sort of *feeling*, the material element
in consciousness. In this signification, thought is
not more immaterial than sense; both are material,
inasmuch as both contain feeling. But what is the
sense in which the terms material and immaterial
are used in the argument now under consideration?
It is this. An "immaterial principle" is described
as "a principle which, although as life-form it makes
perfect another thing, the body, yet is in its existence (*Sein*) independent of this thing."[1] And the
argument is,—Sense is dependent upon the body,
thought is not dependent upon it, except mediately,
so far as it is dependent upon sense. "And because
this activity of the soul" [sense cognition] "is so
bound to the organ of the body that it fully operates
(*sich vollzieht*) in that organ, therefore the sense perception (*Vorstellung*) can be only a material one.
But at the basis of every intellectual cognition, on
the contrary, there lie perceptions which contain
nothing material, and which therefore cannot be
formed in any bodily organ."[2]

This seems to me a string of assumptions. It is
an assumption, that there exists in the body a "principle" independent of the body. It is an assumption
that sense perceptions are themselves material because *produced by means* of material organs. It is
an assumption that there are perceptions differing
from these in being incapable of being formed in any

---

[1] Phil. d. Vorzeit, § 108. Vol. I. p. 187-8.
[2] Work cited, § 28. Vol. I. p. 43.

CERTITUDE AND TRUTH. 191

Book III.
Ch. X.

§ 2.
Modern
Scholasticism.

bodily organ. Indeed it may be said that they are not put forward by the Scholastics in any other character; for they occur, with much more to the same effect, in explanation of the third and last of the three fundamental propositions (*Grundsätze*) with which Father Kleutgen begins the work which we have been discussing. Here then we may expect some light to be thrown on the tenability of the whole Scholastic philosophy.

These three fundamental propositions are:

"First Proposition: Cognition arises by an image of the thing known being produced in the thing knowing, by or from (*vom*) the thing knowing and the thing known."[1]

"Second Proposition: The thing known exists in the thing knowing after the fashion of the thing knowing."[2]

"Third Proposition: Cognition is the more perfect, the farther removed the cognising principle is in point of nature (*in seinem Sein*) from materiality."[3]

Vague and indeterminate assumptions indeed are these, which, like the much more modest *Causa Sui* of Spinoza, require that we should know what is to be understood by them, before we give our assent. The two clauses of the First Proposition are expressly called Axioms[4] by Father Kleutgen; and we thus find that we are expected to assume as axiomatic the separateness of the two things, the knowing and the known, as two independent existents; and the image produced as a third thing intermediate between them. "According to the proposition of Scholasticism which we are considering, cognition presup-

---

[1] Phil. d. Vorzeit, Vol. I. p. 26. [3] Work cited, Vol. I. p. 41.
[2] Work cited, Vol. I. p. 34. [4] Work cited, Vol. I. p. 26.

poses not merely activity on the part of our mind, but also some influence or other on the part of the object."[1] The method of direct perception, as distinguished from reflective, is hereby asserted and adopted. And we see by the plainest evidence, of how little worth is the vaunt of Scholasticism, quoted above, that it is based on reflective cognition.

This assumption of the First Proposition leads at once to a "difficulty," as might have been expected. "But how is it possible that material things should exercise this producing influence on the immaterial mind? This question leads us to the greatest difficulty in the theory of cognition."[2] Just so. Their first "axiom" lands them in their "greatest difficulty." And a very great difficulty it undeniably is, and one which meets the separatists at every turn, in different shapes, according as they shift the *locus* of the meeting between the two concurrent agents; its *locus* I mean in the phenomena of cognition, whether it lies between thing and sense perception, between sense perception and imagination, between imagination and *intellectus possibilis*, between *intellectus possibilis* and *intellectus agens*. At whatever point mind meets matter, these two agencies must act on each other, if they are to produce anything in conjunction; and yet according to Scholasticism nothing material can affect with change an immaterial being;[3] nor can a bodily organ possibly take up into itself an immaterial image.[4]

I think I have now established my point, that

---

[1] Phil. d. Vorzeit, Vol. I. p. 32.   [2] Work cited, Vol. I. p. 33.
[3] Work cited, Vol. I. p. 125. The dictum appears as used by an objector, but appealed to by him as an admitted principle.
[4] Work cited, Vol. I. p. 13.

the fontal error of Scholasticism consists in treating existence as an object of direct instead of reflective perception. It starts with object and subject *separate*, and this involves the absolute, or mutually independent, existence of each of them. This method is common to Scholasticism and to the positive sciences; but it is in place in science, out of place in philosophy. To apply it in philosophy is to make entities of abstractions, to treat the supposed hidden "essences" and "substances" of phenomena as if they were real things, "real grounds," *veræ causæ*; to change philosophy, not indeed into science, but into a science of phantasms.

§ 3. But let us turn in another direction. The error now signalised, though fontal in Scholasticism, is by no means left behind at the source; it affects the whole stream from the source downwards. It reappears in another shape, apart from cognition of self, in the doctrine of Certitude as treated by Father Kleutgen. It is visible in his manner of dealing with certain "truths," which are asserted, and *saving the separatist assumption* truly asserted, to be such as it is immoral to deny.

"That there is such a thing as truth, and not only a truth of the senses, but a supra-sensible actuality; that there is a difference between the morally good and the morally evil; that man must surely hope to attain the happiness for which he thirsts, by loving and practising the morally good; that there is a supreme Originator (*Urheber*) of all things, and through him a moral world-order; these truths, we say, are in such wise revealed to man, in his reasonable nature itself, that it is sinful in him to doubt them."[1]

Let us see how Father Kleutgen deals with an objection which he finds made by the school of Hermes to this view, an objection which he thus states: " It is objected—' that the fact of an involuntary cognition of truth, arising from the reasonable nature of man, is not denied, nor yet that this cognition is accompanied by certitude; only this is maintained, that such a certitude can continue only until the mind begins to investigate the content and the ground of its cognitions, until, as Hermes says, reflection comes in. Then it is that man can no longer rest contented with that cognition which he has, without knowing whence it comes.' " Father Kleutgen answers this objection as follows: " He cannot content himself, because he has a natural need of referring everything that he knows to its ultimate grounds, by investigation,—this we too assert ; he cannot content himself because all and every truth becomes uncertain to him before he has succeeded in referring it to its last grounds,—this it is which we deny."[2]

Observe the method of dealing with the objection; it is the absolutist method, the separatist method. He draws no distinction in the content of the truths themselves, but he denies that those truths, taken as they stand, become uncertain when we have begun to seek their ultimate ground and before we have found it. The truths (he holds) are still certain to us, though we are seeking a reason for them. But surely to seek a reason for anything implies both

[1] Phil. d. Vorzeit, Vol. I. p. 365.
[2] Work cited, Vol. I. p. 369-370.

that we think it wants a reason, and also that we have not found it. Surely this is to be uncertain.

The metaphysician's treatment of the objection would be very different. He would seek to distinguish, in the content of the truths themselves, a part which was immediately and constantly perceived and was therefore self-evident, and a part which was a judgment or a theory attaching to that immediate perception. He would analyse the proposed truth with this purpose, and if it did not at once yield an object of immediate perception which was incapable of further analysis, except into metaphysical elements, he would repeat the analysis until such an ultimate object or objects of immediate perception were arrived at. This object he would maintain as that to which the certitude attached; and would assign to the various judgments, which were combined with it in the proposed truths, a degree of uncertainty greater or less according to the grounds on which they might respectively be supported. The metaphysical method is analytical.

Scholasticism on the other hand will have its truths in the lump, *in globo*, to borrow an expression from Dr. Newman. And why? Because the judgments, which are combined with the ultimate and immediately perceived certainties, make those certainties. into objects of direct perception, or absolute existents; and furnish the enquirers by this means with finite bits of truth, so to speak, existents at once real and true in philosophy, and yet accommodated to the popular common sense, which knows nothing of reflective or philosophical perception. The world, its creator, the moral order, the soul of man, free will, and so on, are instances of such supposed objects of direct perception, some of them sensible, others supra-sensible: but all alike formed by clumping judgments with immediate percepts, and then claiming for the compound the inviolate certitude which in truth attaches only to its perceptual elements.

By this way of clumping a perception with a judgment arises that confusing mode of speech which attributes truth as well as reality to *things.* Truth and falsity are attributes only of *judgments;* reality and unreality of things, that is, of percepts. And this is the Aristotelian doctrine. On the other hand let us see how the Scholastics, while recognising this doctrine, yet manage to obliterate it and attribute truth to things. "Every thing (*Sache*) which exists is also we must admit (*freilich*) true; but it is true only by having reference to cognition, that is to say, that it is or can be cognised. Just so also is every true thing (*Sache*) evident; but it is so by its truth becoming manifest."[1] Here truth is claimed as an attribute of things, notwithstanding that it is only made manifest by cognition, that is, by a process which includes judgment. And the claim is put forward as if it were a *concession*, in the word *freilich*. But be it claim or concession, it cannot be granted in the full sense in which it is demanded. It is not the case that every existent thing is *true;* for, since "it is true only in reference to cognition," its *permanence* in thought, and not its existence merely, is what its truth consists in. Truth is not a quality or attribute of its *qua* existent. But more on this point farther on.

By this way of taking the matter, the step is already made from things as known to things-in-

[1] Phil. d. Vorzeit, Vol. I. p. 434.

themselves, to absolute and separate existents. For judgment is thereby made into a perceptive instead of a conceptive operation, that is, into an operation which perceives, not the connection between percepts, but the inner nature of a single percept. Thus we read: " But just in this consists" [according to St. Thomas] " the difference between intellectual cognition and sense cognition, that the latter busies itself with certain external qualities, but the former presses on into the essence (*Wesen*) of the thing. For this essence is, even according to Aristotle, the appropriate (*eigenthümliche*) object of the Reason. And what the sainted teacher here names the essence of the thing, up to which he says intellectual cognition presses forward, that he directly afterwards designates as the *hidden nature of a substantial thing* (*Dinges*), hidden under the phenomena."[1] In other words more calculated to show the fallacy,—because intellectual cognition is a process which aims at discovering the nature and laws of phenomena, and can be characterised as having that for its proper object, therefore *there exists* a supersensible kernel in sensible objects, which supersensible kernel the intellect actually perceives. And yet this is not our old acquaintance, Intellectual Intuition?

> "Jane Lamb, that we danced with at Vichy !
> What, is not she Jane ? Then, who is she ?"

§ 4. We are now prepared for the examination of that point in scholasticism which is the key of all the rest, which furnishes us with the basis of its

---

[1] Phil. d. Vorzeit, Vol. I. p. 151-2.

doctrine of certitude, and of its doctrine of self-cognition, I mean its general doctrine of cognition. Here if anywhere is to be sought the reason which justifies the scholastic notion, that the "substance," which they maintain that we cognise intellectually, is a real existent. This doctrine of cognition is expounded in the second and fourth Chapters of the first Division of Father Kleutgen's work; its fundamental traits are given in the former, and its method delineated in the latter, and chiefly in the first Section of the latter chapter, headed the Nature of Abstraction (*Begriff der Abstraction*).

The second Chapter follows immediately upon the exposition of those three fundamental Propositions already cited, and begins as follows: "Not merely outside of objects of sense do we think beings which are not perceived by sense, *e.g.*, pure minds (*Geister*), but we think also, in the objects of sense themselves, something which is not perceivable by sense. For everything which we cognise by the senses, *e.g.*, hardness, odour, sweetness, colour, shape, we think not as the thing, but as something in the thing, as that whereby the thing becomes perceivable, or is manifested to us. Though therefore all that is accessible to sense should undergo manifold change, the mind ceases not on that account to recognise the thing as the same. But just for this reason that the mind apprehends the different impressions, which sense receives, as consequences of a change which the thing undergoes, it is evident that it" [the mind] "thinks something else (*noch etwas anders*) of or rather in the thing than what sense perceives, namely, a permanent something which is changed, and accordingly distinguishes from the changeable *phenomena* (or *accidents*) the *essence* (or the *substance*)."[1]

Three remarks on this very striking passage. First, does it mean to assert that the substance is changeable or unchangeable? What is the meaning of "a permanent something which is changed, *ein Bleibendes das verändert wird*"? If it is not changed, what connection can it have with the change in the accidents? If it is changed, why is it contradistinguished from them as "permanent"?

Secondly, observe the assertory and subjective evidence for this substance, "*wir denken*," *we think it.* Suppose we do; suppose this is a correct account of what we actually do think; still, does this show that we are *right* in thinking this mass of contradiction? that, in spite of the contradiction, the *permanent substance*, which is part of it, is a *true* thought?

Thirdly, observe the tacit assumption with which it begins, namely, that phenomena come before us at and from the first ready marked out into separate concrete things. We are supposed to find ourselves in a world of separate objects, to observe their changes, and to compare them with each other. Their separation is a *datum.*

But now let us take another passage from the same page. "The properties, states, phenomena, of a thing can undergo change of many kinds, without our having to suppose on that account a change of its substance (*Wesen*); we think this also still the same. Just so we can perceive difference of many kinds in many individual things, and yet think these things by means of a representation (*Vorstellung*), by means of that representation, I mean, which appre-

---

[1] Phil. d. Vorzeit, Vol. I. p. 53-54.

hends only the essence (*Wesen*) common to all the individual things. These representations therefore are *universal* (*allgemeine*), i.e., such that their object can be multiplied in actuality, or what is the same thing, their content (*Inhalt*) can exist in many things."[1]

Observe the theory; the substances of things of the same class are all precisely alike (so far as our thought is described as having gone at present), only different *numero;* but the representation, the *Vorstellung,* by which we think them, is *one* as well as the same, is an universal applicable to all the precisely similar substances. We think by forming universal representations in the mind, each of which corresponds and is applicable to any number of substances, all precisely alike, or, if different, yet with a difference which has as yet escaped our notice.

The exposition proceeds to describe the process of thought in the other branch of the case, that is, not of the substances in one class, but of the phenomenal properties and accidents in a substance. " But not merely substances, which only reason cognises, but also all those properties, states, phenomena of these substances, which sense perceives, are thought by the mind by means of universal representations (*Vorstellungen*). These too the mind divides by means of these representations into genera and species, and in them too distinguishes accordingly the essential from the accidental."[2]

Observe the leap in the theory of substances. They suddenly appear as *cognised by reason,* " *nur die Vernunft erkennt.*" Before, it was " *we think it*" only. Their existence poses now as an indisputable cogni-

[1] Phil. d. Vorzeit, Vol. I. p. 54.   [2] Same passage continued.

CERTITUDE AND TRUTH. 201

Book III.
Ch. X.

§ 4.
Cognition in
Scholasticism.

tion. Nor is this merely a passing inadvertence. We soon find it said, without any further reason but a description of sense and imagination processes, " According to this, the intellectual representations differ from the sensible in this, that the senses always perceive only what is individual (*einzelnes*), and this only according to its *external phenomena*, but the reason thinks beside the phenomena the *essence* (*Wesen*) also, and this, as well as those, by *universal* representations."[1]

Such being a general sketch of intellectual cognition, the question occurs, in what its distinctive nature is supposed by scholasticism to consist. Is it in the power of forming universal *conceptions* in the mind simply? Or is it in the getting to the individual *substances*, several in number, embraced by a single universal? Or is it finally in the insight, that the individual substances really exist and correspond to the universal conception? If the second or the third, some proof of the *fact* would be desirable.

But on the contrary, the farther we read, the wider becomes the cleft between the universal and the substance ; the *Wesen* which is thought and the *Substanz* which exists. " The essence (*Wesen*) of the individual thing includes indeed no attributes but those of the species; nevertheless something must be thought in it, without which it cannot be, and which can be in no other thing; that is to say, just that whereby it gets its individual determination. There is therefore in the individual things in any case (*allerdings*) something, which is essential to them as individual things.

" But do we cognise this? We cognise it at any

[1] Phil. d. Vorzeit, Vol. I. p. 55-56.

rate (*freilich*) so far as we see that a thing can indeed have its constitution (*Beschaffenheit*) in common with others, and be perfectly like them; but that it has necessarily an existence (*Sein*) peculiar to itself, and common to no other thing."[1] Then how can this existence be cognised *by an universal?* * * *
"In any case however we understand that things could not be different from one another in number, if they had not, each of them, their own peculiar existence (*Sein*), and had not, in it, something essential (*wesentliches*) to them, which the concept of the species does not include in itself."[2]

Here then the *substance* escapes us altogether. We seemed to get hold of it originally, because we thought we knew something *about it*, that it was a permanent substrate and so on; but now it appears that we cannot grasp it at all; even the universals fail us and fall short of the substance, which is different in its content, as it now appears, and not only *numero*, in every individual thing. The substance was an individual object of thought, and not of sense; an individual thing thought, together with others, by an universal concept;—but now it appears that by this universal concept we cannot think individual substances at all, we can only think the *species*. These substances then,—what becomes of them? We cannot *see* them, and we cannot *think* them; they exist neither for sense nor for thought. And we seem in danger of falling into Scotus Erigena's way of thinking. "invenies *οὐσίαν* omnino in omnibus, quae sunt, per se ipsam incomprehensibilem non solum sensui, sed etiam intellectui esse."[3]

[1] Phil. d. Vorzeit, Vol. I. p. 57.   [2] Work cited, Vol. I. p. 58.
[3] De Divis. Nat. Lib. I. § 25.

We can *imagine* the individual substances, however, from analogy with the concrete individual things of sensible experience; and the imagination, though groundless, is not unattractive to some minds. It should be noticed, too, that the word *substance* is dropped in the passage last quoted, and replaced by the more vague and modest *Sein*. This, however, may be in order to embrace under one term the two cases of reasoning, namely, reasoning about the nature and laws of substantial things, and reasoning about the nature and laws of attributes apart from substances.

I pass now to the exposition of the method by which the reasoning about substances is instituted and carried on; and I turn to the first division of the fourth Chapter, headed *Begriff der Abstraction*.[1] Cognition by abstraction is the name by which the scholastic theory designates the process.

This process of cognition by abstraction is carefully distinguished from that of Locke, which is said to be an abstraction by means of comparison of sensible phenomena, beginning from the most special and rising to the most abstract and general determinations. "So one argues against Locke, and supposing that the Scholastics also hold abstraction to take place by comparison of many sensible perceptions (*Vorstellungen*), one believes that they, too, can be refuted in the same manner."[2]

The abstraction of the Scholastics is a severing, in thought, of the essential from the contingent; the reason thinks an individual thing, which sense perceives, thinks it *e.g.*, as a stone, or as a tree; con-

[1] Phil. d. Vorzeit, Vol. I. p. 109-118.
[2] Work cited, Vol. I. p. 114.

siders in it only that which is common to all stones or trees, and which on that very account is, in each stone or tree, the law by which it is determined. This is the essentiality, *Prinzip*, or determining ground, of its manifested existence (*Dasein*), and is always thought as *in* the stone or tree, not outside it, as Plato is said to have held; "and the apprehension of the individual thing by the abstract concept is on that account, though indeed imperfect, yet not untrue."[1]

This is then farther explained by distinguishing three degrees of abstraction; 1st, abstraction merely from the accidents of the individual thing as *this* individual, retaining the properties which are common to it with other bodies and are perceived by sense; 2nd, abstraction from sense perceptions, retaining only the "first attributes of all bodies," magnitude, extension, and shape; 3rd, abstraction from what belongs to bodies as such, retaining only that by which non-bodies can be apprehended also, (*Vorstellungen durch welche auch Unkörperliches aufgefasst werden kann*). "These are the highest concepts, existence, unity, substance, accident, faculty (*Vermögen*), activity, force, and so on. We have already remarked (No. 28. 63.) that no intellectual representation whatever takes effect without these highest concepts, and for this reason too the degrees of abstraction, which have been mentioned, must not be so understood as if, in forming concepts, we began with the first and proceeded to the second and third. Rather it is just the highest concepts that are the first by which the reason thinks every kind of object."[2]

---

[1] Phil. d. Vorzeit, Vol. I. p. 110.

[2] Work cited, Vol. I. p. 111.

"We form, for instance, the concept of man, by thinking him first as being (*Wesen*) or thing, then as living, then as sensitive, and finally as sensitive and reasonable being."[1]

Of course the question must occur to every one, how come we in the first instance to apply the concept of existence (*Sein*), taking that as the most abstract and highest of all? Where do we get the notion from originally? How does it *arise?* The scholastic answer is ready;—we have a *Faculty of Reason.* The concepts come directly from that faculty, when applied to sense perceptions, (which, as I have already remarked, are supposed to be at and from the first ready made up into separate things). "According to St. Thomas, the highest and simplest concepts are the first which we form."[2] To think is to form concepts, and we have the faculty of thought. And the method is to proceed from the simplest to the more complex. "Therefore we heard the Scholastics say, at the very outset of the first explanation of abstraction, that, as the eye perceives colours in the object, and the ear sounds, so the reason also apprehends in it that which is proper to it to cognise, the nature and essence. If now it is added to this, that the representation whereby we think the nature of the object is an universal one (*allgemeine*), still it is not meant by this, that we first seek in many things for what is common to them all, and think this as the essence of the individual thing, but that the thought (*Gedanke*), with which we apprehend the essence of the individual thing, at first includes only that which is common to it with many other things."[3]

[1] Phil. d. Vorzeit, Vol. I. p. 112.
[2] Work cited, Vol. I. p. 114.   [3] Work cited, Vol. I. p. 115.

The comparison of reason to the eye and the ear makes us think of it as *perceiving* its " proper" object, the *substance*, by a sort of intellectual intuition. But this doctrine is, as we have seen, expressly repudiated, and it is claimed as a *conceiving* faculty or process, a *thinking* by general terms. Now I think the Scholastics of St. Thomas' school are in this dilemma, —either they must maintain that reason is a conceiving and not a perceiving faculty, and then they have no proof of the existence of a *substance*, since that substance is *individual* in each concrete individual thing, while reason can give only *general* terms; or they must admit that reason is a *perceiving* faculty and has a direct perception of the individual substances in the phenomenal things. And I suspect that it was chiefly the apprehension of this inevitable dilemma, in that remnant of scholasticism, the Leibniz-Wolfian philosophy, which drove Kant to his assumption of the Transcendental Categories of the understanding, as the sole means open to him of explaining the transition from the individual in perception to the general in thought, and so saving the reality of empirical objects.

If this view is correct, the only issue for the Scholastics from the dilemma I have proposed to them would be that issue in which they have been preceded by Kant, and in his case found impracticable, namely, the assumption of *a priori* and transcendental forms of thought, to which their list of highest and simplest concepts just quoted would belong. Theirs is a case of arrested development. They have still to become in succession Cartesians, Leibnizians, and Kantians; then they will stand upon the threshold of philosophy.

§ 5. I pass in the next place to the scholastic theory of Truth, and to the relation between its objective and subjective sources. The question is pre-eminently a metaphysical one; its object-matter, truth, involving immediately and inevitably the two inseparable aspects, objective and subjective. Now what the Scholastics begin with *separating* they must find some means or other of *combining;* they cannot *end* with things there, thoughts here, or they would have no philosophy at all. Accordingly, they conceive Truth (which it is the purpose of philosophy to discover) as consisting in the union of the two, the agreement of thought with thing. The question is, how does this union of separates take place, and is its product, truth, a thought or a thing? They have to ride, as it were, two horses round the ring at once. At one time truth appears "subjective," at another "objective;" at one time it seems to exist because *we cognise*, since it is found originally only in judgments, not in single terms; at another because we cognise *the thing*.

Two passages from Father Kleutgen will show us the difficulty and its proposed solution. First the difficulty. "Yet how do we get to the notions (*Vorstellungen*) of the *True* and the *Good?* By the consciousness of our own cognising and desiring, answers St. Thomas, and we may add that we get to the idea of the *Beautiful* through this same consciousness of what we experience in ourselves. Truth is primarily (*zunächst*) a property of the cognising (*des Erkennens*), but a property the notion (*Vorstellung*) of which is obvious (*sich ergibt*) to us at once from the nature of cognising. Cognition (*die Erkenntniss*) is true by its corresponding to the thing (*Sache*).

But what else is the cognising process itself but a mental seizing, an intellectual possession of the thing? But we cannot apprehend it as such without cognising as its necessary property, that it agrees with the thing, and therefore is true. But when we think the true as something permanently existent (*Bestehendes*) out of us, and say, for instance, that we seek truth or know the true, then that first notion lies at the root of this conception. For we understand by this Truth, which we seek, nothing else than a thought or a proposition to which the actuality corresponds."[1] The nature of cognition, of the cognising operation itself, is to take possession of the thing. Cognition makes its own what was before not its own. Is, then, truth inseparable from cognition; are all cognitions true, or are some erroneous? And what is the criterion discriminating truth from error, or real from apparent cognitions? How do we know when we have really got *possession of the thing?* or that it is *the right thing*, of which we have got possession? A further question is thus opened by the scholastic account of truth, a question of analysis of the relation of thought to thing in cognitions that are true; and it is on this point that light is required.

But it is just here that the scholastic explanation fails us. Instead of giving us such an analysis, they give us a theory of the *genesis* of truth, *assuming* that the relation between thought and thing is a causal one, and assuming also that truth in cognition depends on something external to the cognition. If we turn back to the remarks explanatory of the first of those three fundamental propositions, which have

[1] Phil. d. Vorzeit, Vol. I. p. 107-108.

been already quoted, we shall find a general view of the nature of truth, which will bear me out in this statement. We there read: "Just as there is in all things a conation (*Trieb*) not only to preserve but also to spread and propagate their existence (*Sein*), so also there is in all a striving and a faculty to manifest themselves, to exist in others by this manifestation, and so to propagate their ideal existence (*Sein*). In this faculty (*Vermögen*) of manifesting themselves to the mind we may therefore find a cause (*Grund*) why all things not only are, but also are *true*. That is to say, truth is predicated of the existent on account of its relation to the cognition-power; things therefore are true because in their actuality they correspond to the ideas of the creative mind; but they are also true, because they are adapted (*geeignet*) to produce in the created human mind a cognition of themselves."[1]

This passage must not, of course, be regarded as an *answer* to the questions which I have raised above. No. I bring it forward rather as containing the conception which prevents them from being fairly raised at all. For thereby the analytical question is shelved, and replaced by a question of genesis, by introducing the distinction between creator and created. But the introduction of a question of genesis is equivalent to and involves the assumption, that truth in cognition depends on something external to the cognition. The question *how* we can lay hold in thought on anything wholly external to thought is replaced by an enumeration of two cases or instances, which exemplify the supposed laying hold in thought; first, our thoughts being true when they correspond to things; second,

[1] Phil. d. Vorzeit, Vol. I. p. 33.

things being true when they correspond to thoughts in the divine mind.

We might also proceed to another instance, on the principle of *operari sequitur esse*, and say that the thoughts in the divine mind are true because they correspond to the divine mind. We should then have the absolutist chain of truth complete. *We* know things because they first are ; we know ourselves because we first are ; God knows Himself because He first is. Observe too the occurrence, in the passage quoted, of the notion of a self-manifesting faculty ; a notion which we have already found in several modern absolutists.[1] who urge that the source of existence is a power of self-manifestation.

The covert passing over from the order of subjective analysis to that of empirical and objective genesis, which is effected by simply introducing the distinction of creator and created, is the logical machinery by which the absolutist theory of the nature of truth is erected. The metaphysician's task must be to bring back the question to its true arena, by referring it to the tribunal of reflection ; and this it is which I have attempted to do in previous Chapters, as well as in the present criticism of the opposite theory.

Answer, properly speaking, to the questions I have raised there is none ; there is nothing but a repetition of the same views of cognition again, made to serve in place of one. The whole third Division of Father Kleutgen's work is devoted to *Certitude*. And we find there at some length the distinction between actual and methodic doubt; that between objective and subjective ground of certitude; that

[1] Above, Chap. VII. p. 27 et seqq.

between its Rule (*Norm*) and its Ground (*Grund*); and the question discussed and answered, how far the human reason is infallible (*untrüglich*). But the whole result is no more than what we have already seen. The light that dwells in every man, "dieses *jedem* Menschen *inwohnende* Licht," is the individual Reason. And this reason is at once the rule (*Norm*) and the ground (*Grund*) of the certitude in our rational cognition, because the reason can make itself and its cognition an object of its own consideration.[1] And this also sets limits to certitude, while securing certitude within them. "Reason, says St. Thomas, cannot err in those judgments which are cognised from themselves, and thence comes the infallibility (*Untrüglichkeit*) of that which is inferred with certitude from first principles."[2] All which would be undeniable, if only (and here is my objection), if only a particular and erroneous theory of cognition was not assumed to be spoken of in speaking of cognition; if only a truly reflective act, and not a direct act *called* reflective, was intended by the term self-cognition.

We have now before us the whole process of scholasticism in its reasoning on the question of truth as the ground of certitude. We possess certitude, it is argued, because we know that we know; and we know that we know, because we know *ourself* as the ground of our own knowing. This is the argument from self-consciousness. But secondly, how do we know *ourself?* We know ourself as we know other "substances," by an intellectual cognition the property of which is to cognise the "essence" of indi-

---

[1] Phil. d. Vorzeit, Vol. I. p. 452. 453.
[2] Work cited, Vol. I. p. 455.

vidual things. Thus the theory exhibits self-consciousness, or reflection, as a mode of direct consciousness, and makes it depend upon the direct cognition of the substance in the phenomena cognised. But this substance, as I have shown, is a pure assumption; therefore so is its cognition; and the assumption is made by adopting the direct method, which clumps judgment with perception, instead of the reflective, in the case of objects not-ourselves; and in the case of self, by clumping the reflective and the direct methods together, that is, by assuming a substantial *self* as the object cognised.

The basis of certitude is thus laid in the cognition of self; and the basis of the cognition of self in the cognition of objects generally. But the cognition of objects generally is the cognition of their "substance," the faculty of intellectual cognition being of such a nature as to know the "essence" of things. And the ultimate assumption at the basis of the whole fabric is this, that we have such a faculty. This no doubt is what is meant by the claim of scholasticism, that it supplies a "guarantee of actuality," (*eine Bürgschaft für die Wirklichkeit*).[1] It proves the reality of presentative percepts by connecting them as phenomena with the noumenal "substances" which the intellect cognises. The supposed *cognition of noumenal substances* is the real basis of the scholastic philosophy.

But what an empty notion is this, that presentative percepts require a guarantee! It is they that are the guarantee of everything else, representations, concepts, ideas, theories, philosophies. For all must

[1] Phil. d. Vorzeit, Vol. I. p. 467.

not only start from presentations, but also be verified by them, if verified at all. Theories do not verify facts, but facts theories. How groundless is the fear of making philosophy unreal, or of not securing it against being so, by giving up the notion of Absolute Existents behind phenomena. As if the absence of this notion could make philosophy unreal; as if (supposing there was not such an existence) our holding that there was could remedy the evil, and make philosophy real again. And in what, after all, does the proposed guarantee for the actual consist? In the assumption of *another actuality* behind it.

Consider again too, in this connection, the scholastic definition of Truth, which is also the current definition, the definition of popular philosophy, being that of the direct mode of consciousness as distinguished from the reflective. That definition is—the agreement of our thought of things with the things themselves. But the question is and always must be,—How do we know things *except* by thought? Answer, We do not. How then can we know that they agree with thought? Answer, Only so far as one thought agrees with another. Hence the definition of Truth which modern philosophy proposes;— the agreement of thought with itself, after the most complete examination of its content. This provides for the verification of representations, and of theories of any kind, by presentations, since that is included in the term examination. And it is a definition which embraces truth both scientific and philosophical.

Observe, too, the difference of *place* which truth holds in philosophy, according as it is defined in this way or in that. In the definition given by modern philosophy, truth is an *ideal*, a state of knowledge which is the *goal* of enquiry. In its scholastic definition, it is the *starting point*, being assumed to be known at first and always by intellectual cognition, which is a faculty for cognising "essences." Not that truth is represented by scholasticism as perfect or complete from the first, any more than it is by modern philosophy; but that it starts with a particular piece of truth ready made in every case, namely, that its object is a "substance;" and although there is room for indefinite perfection in the knowledge attainable of the nature of substances, yet the notion of substance can never be transcended or got rid of, and is the object of truth at the end of all enquiry, as it was at the beginning. We have, in short, by this view of truth and this assumption of substances, the universe broken up into particular existents; for every substance is a particular existent, and such existents are the ultimate components of the universe, and known to be so on the scholastic theory of cognition.

§ 6. The full exhibition of the Scholastic theory must be sought in systematic works such as that of Father Kleutgen, which I have now been criticising. From this it has been seen, that the doctrine of Certitude depends upon the doctrine of cognition generally, through the medium of cognition of self. Nevertheless, as already noticed, we sometimes find that a doctrine of Certitude is put forward as having a basis of its own, carrying its own evidence along with it, without any reference, explicit at least, to the theory

CERTITUDE AND TRUTH. 215

Book III.
Ch. X.

§ 6.
Intuitive
Certitude.

of cognition generally. This is the case in Mr.
Ward's Philosophical Introduction to his Nature and
Grace, already cited. Whether his theory of Certitude by Intuition is compatible or not with Father
Kleutgen's exposition of the scholastic theory of
"cognition by abstraction," I will not attempt to
decide. But it will be clear, I think, to any reader
of it, that it stands on a basis quite independent of
that theory.

It is a real pleasure to read Mr. Ward's writings,
not only from the perfect lucidity of their style, but
from the conviction which they inspire of the perfect
sincerity of the writer. It is with Mr. Ward's writings
as it is with Dr. Newman's; they are writings that
kindle the desire of agreeing with them; and they
leave behind them the keenest regret that agreement
should be impossible. I only hope that my argument
against Mr. Ward's theory may catch some breath of
that spirit of kindliness and courtesy which pervades
his exposition of it.

The Section which I am about to criticise is
entitled "On Intuitions and on the Principles of Certitude."[1] It begins by distinguishing "judgments of
consciousness" from "judgments of intuition." Judgments of consciousness express "the mind's reflection
on its own actually present experience;" *e.g.*, that I
am now feeling cold. Judgments of intuition are
more than this, but yet not inferential. Four instances are given of them. 1. I *remember* that, say
half an hour ago, I felt cold; 2. My judgment that *if*

---

[1] On Nature and Grace. Philosophical Introduction. Chapter I.
Sect. I. p. 5. Parts of this same Section have been already criticised in Chapter IV. Vol. I. pp. 254 and 274, to which I would
beg the reader to refer.

the premisses are true, then the conclusion is true; 3. Mathematical Axioms; 4. My judgment that there *are* external objects. Such *intuitive* judgments, it is said, may be true or may be false; it is as *immediate* and yet not inferential, *immediate* and yet more than judgments of consciousness, that they are called intuitive. So much for the explanation of the term.[1]

We now come to the question as to the truth or falsity of such intuitive judgments. The sceptical position is stated to be, that "no intuitive judgment can possibly carry with it its own evidence of truth."[2] The contradictory of this position is what Mr. Ward calls the *Principle of Certitude*, namely: "'it is fully possible that intuitive judgments may carry with them their own evidence of truth.' And this proposition may well be called the Principle of Certitude; because, unless we confidently maintain it, it will be impossible consistently to recognise the certainty (or even approximation to certainty) of any one thing, beyond our actually present experience. If this principle were untrue, our knowledge would be less than that of the brutes; it would be strictly confined to the mind's reflection at each instant on its own existing consciousness. We could not compare *e.g.* our present consciousness with our past; for unless the Principle of Certitude were true, we could not even guess what our past consciousness has been. Much less, as is evident, could we even contemplate comparing our own consciousness with that of others."[3]

---

[1] On Nature and Grace, p. 5-7. And again p. 39-40.

[2] Work cited, p. 9.

[3] Work cited, p. 18. And see this expanded and enforced from p. 26 to 33.

CERTITUDE AND TRUTH. 217

BOOK III.
CH. X.

§ 6.
Intuitive
Certitude.

Now what reply is there to this argument, that there may be judgments which are intuitions of truth? In the first place I reply—Observe the term *truth* in the statement of the principle. It is not expressly defined; but it is abundantly clear from expressions which precede it, that it is intended to adopt the usual definition, 'agreement of thought with thing,' without question asked,—what are things *except* as thought. The ambiguity of the term, whether latent in the definition or in the word, covers the assumption of a real, separately objective, existence in the object-matter of the judgments which are recognised as true. Thus in the explanations which Mr. Ward gives of his four instances of intuitive judgments, what he urges is, that in these cases we do not merely say 'I am impressed with the feeling;' 'I cannot help thinking;' but we state confidently 'the proposition *is* true;' the judgment, he argues, in these cases is not *subjective* but *objective*, not a judgment merely of consciousness, but an *intuitive* judgment.[1] Here we have the whole consequence of the ambiguity in the. definition of truth. Because our judgments of memory, of connection between premisses and conclusion, of axioms, of external existence, are as confident and unhesitating as judgments of consciousness themselves (*e.g.* that I am now feeling cold), therefore these judgments have an *objective* character as opposed to a *subjective* one. Judgments of consciousness are now spoken of with a 'merely' before them, 'mere judgments of consciousness,' as if implying that there was more *objectivity* in the intuitive kind than in them. Now what there is more in them is the *representation* which is included

---

[1] These expressions are used p. 6. 7.

in their content, whereas judgments of consciousness include only *presentation*, or presentative perceptions, *e.g.*, an actual feeling of cold. There is more in them than in judgments of consciousness, because they include a representation as well as a presentation; they have not more objectivity than judgments of consciousness.

It may be well to observe, that Mr. Ward in the passage now under examination does not *define* what he means by judgments of intuition; he gives instances, he opposes them to inferences, and to judgments of consciousness; but he leaves it undecided whether he would oppose them also, in definition, to *subjective* judgments, and whether he would accordingly include *subjective* as part of the definition of judgments of consciousness. So much by way of preliminary objection. I now proceed to the substance of my reply, which however will consist in nothing else than applying the distinction which I have just drawn.

In fact, if we studiously abstain from involving the notion of what Mr. Ward calls *objective*, but I absolute or separately-objective, existence with the judgments which he calls intuitive, then his distinction between judgments of consciousness and judgments of intuition is nothing more nor less than a distinction between judgments, the object-matter of which, as seen in reflection, consists the one of presentative the other of representative perceptions. 'I am now feeling cold' is a judgment, the object-matter of which is a presentation. 'I remember to have felt cold half an hour ago' is a judgment which has representations for its object-matter.

Now I never heard of any sensible person, I mean

by this to exclude the sceptic of romance or by pro-
fession, who denied that we *could have* true representations, who limited our possible knowledge of
truth to presentations alone. The fact is that we
instinctively trust our representations, and not those
of memory only, but also of imagination. We trust
them instinctively, subject to verification, to the control of facts, that is of presentations. We do not, as
Mr. Ward supposes, trust them *absolutely*, because we
soon learn that the unquestioning trust with which
we start is not to be relied on. But we trust them
instinctively, though provisionally.

Looking at the phenomena of the case, a few
words will suffice to show how the control and verification are exercised. I think I can show that it is
quite unnecessary to suppose with Mr. Ward, that,
unless the judgments called intuitive carried with
them their own evidence of *truth*, it would be impossible even to compare our present consciousness with
our past, impossible to guess what our past consciousness has been. Mr. Ward proceeds on the tacit
assumption that experience comes to us in separate
portions, called present moments; any one of these
once gone into the past, we should have, he
thinks, no means of knowing anything about it,
had we not an *intuition* of it, that is, a perception
of it, though absent, as if present. The necessity for this violent hypothesis is removed, when
the continuity of perception, as well as its discrete-
ness, is understood, according to the analysis in
Chapter IV.

The memory tells us, say, that we were cold half
an hour ago. The question is, Were we? We have
a present impression to that effect, but may not this

be a delusion? Mr. Ward says,—No; it cannot;[1] it carries with it intuitive evidence of its truth. This simplifies the matter very much for any one who has reason to believe in the infallibility of his own memory. But what is the fact? In what does the present impression of having been cold half an hour ago consist? The content of the feeling itself, *cold*, is the same as in presentation. only much less vivid. The surroundings of the feeling are reproduced with the feeling itself, also less vivid than they would be in presentation. The change in the feeling and the change in its surroundings, from half an hour ago up to the present moment, when we are feeling presentatively warm, are reproduced also. In this way the representation of former cold is shown to be *continuous* with the presentation of present warmth. This is one part of the verification.

The second part is, to see whether my present representation of having been cold is combined with other representations, upon the repetition of which, as to their content, in presentation, the representation of cold also becomes a presentation. The question is—Was I really cold half an hour ago, or is it only my hallucination now? Well, how do I represent the circumstances attending that representation of cold; do I represent myself as having been out of doors without a coat on a winter's day, or as sitting by a warm fire? Can I repeat those circumstances again, go out or sit by the fire, and feel cold again now at this moment? If I remember or represent, along with the cold, the circumstance of having been out of doors, if I can go out of doors and feel cold

—————

[1] This seems intended by Mr. Ward as an instance of a *true* intuitive judgment, p. 9.

again,—then my present impression of having been cold half an hour ago is probably *true* and not an hallucination.

This is a sketch of the mode in which the verification of representations is performed. Presentations are the instrument, and bringing back the *content* of the representations into presentation again is the method. But presentations and representations are alike, completely alike, in being both equally subjective and objective at once. One is not more subjective, or more objective, than the other. A presentative feeling of cold is a fact as well as a feeling; a representation of cold, of having been cold once, is a present fact representing a past fact, that is, a fact in different context from the present; and just as much a present feeling representing a past feeling.

What then becomes of intuitive judgments which carry their own evidence with them? It is clear that they are a pure fiction in cases of simple memory, Mr. Ward's first class of instances. The same might easily be shown with regard to his third and fourth cases, the axioms of mathematic, and the judgment that 'there *are* external objects,' so far at least as this judgment means to assert an absolute or separately-objective existence. But the second class, the judgments which lie at the root of reasoning, may seem perhaps to demand a further notice.

In fact these judgments differ from the verified and therefore now certain judgments of representation, by the circumstance that they really are, what Mr. Ward ascribes to the rest, not demonstrable but self-evident; still, primarily, self-evident only as means of reasoning, not as attributes of percepts; and for this reason are called, as above explained, the Postulates, not Axioms, of Logic.[1] There is no *intuitive* certainty about them, in Mr. Ward's sense of the term. The Postulates of Logic and the syllogistic canons founded on them are parts of method, not parts of phenomena. They express the *intention* which the mind has of holding fast in representation a certain content and of dealing with it in a certain way. They mark our hold upon the flux of consciousness, which they enable us to arrest and examine, and that in portions of it so minute as to render the changes within them imperceptible to presentative sensation. And verification by presentation attests the truth of reasonings performed in this manner, so far as sense permits verification to be performed. There is no separately-objective fact in the syllogism, of which the syllogistic method gives us the intuition. When I say 'if the premisses are true, then the conclusion is true,' I ascribe no separately-objective existence to the syllogism; what I mean is, that if I can verify the premisses, then I know that I should be able to verify the conclusion, under suitable circumstances of experiment.

It has now, I think, been made clear, that the fundamental error in Mr. Ward's theory of intuitive judgments is the same as that in Father Kleutgen's scholasticism, and that this error is no other than the separation of object from subject by adopting the direct instead of the reflective method. This error has been detected in the origination of the theory of intuitive judgments. I proceed to trace it in some of its consequences.

Having established, as he thinks, the truth of the

[1] See above, p. 103.

Principle of Certitude, Mr. Ward proceeds to signalise as the first question to meet us,—What is *the test of legitimate intuitions?*[1] There must be some primary intuitive judgments, which are not only true but the basis of others. Mr. Ward coins the word *intuem*, to express the thing *intued* in an intuitive judgment.[2] The question, then, is,—What intuems are necessary? "Of intuems, some are 'necessary,' others not so. By 'necessary intuem' is meant a verity legitimately intued as 'necessarily' true. When *e.g.* I intue by memory that five minutes ago I was seated at this table, this truth is no necessary intuem. But when I intue that a rectilineal figure of three sides has three angles, the truth is necessary, and is legitimately intued as such."[3]

Mr. Ward does not propose any theory of his own, either as to the test of legitimacy of intuems, or as to the distinction of necessary intuems from the rest. But he proceeds forthwith to consider "the relation in which Necessary Truth stands to the Intellect and the Will" of God; taking for his instance of necessary truth the axioms and theorems of geometry.[4]

Two propositions are implied, says Mr. Ward, in this word 'necessary;' " (1) Necessary truths do not derive their verity from the fact that *God necessarily intues them.* Rather the very opposite is the fact: God necessarily intues them, *because* they are necessary truths. Who would say *e.g.* that God is necessarily Self-existent, *because* He intues Himself to be so? On the contrary of course: He intues Himself to be so, *because* He *is* so. * *.* It follows, therefore,

---

[1] On Nature and Grace, p. 34. [2] Work cited, p. 40.
[3] Work cited, p. 42. [4] Work cited, p. 42-43.

that through all Eternity God is constantly gazing on the vast mass of Necessary Mathematical Truth. (2) It is not only certain that necessary truths are not derived from God's intuition of them, but that in some sense they seem to limit His Power. God cannot create a rectilineal figure of three sides, which has more or less than three angles ; or again, whose angles taken together amount to either more or less than two right angles."[1]

"Here then," Mr. Ward continues, " we are brought face to face with the difficulties which I mentioned as existing."[2] Yes, no doubt; scholasticism could not breathe but in an atmosphere of self-created difficulties. But what is the solution? Necessary Truth *is* God. " This then is our conclusion. All Necessary Truth is identical with God; in intuing it, I really intue the One Necessary Ens; though in this, as in so many other cases, I may be very far from recognising the full extent of the Object which I contemplate. God, in intuing Necessary Truth, intues Himself; a creature, in intuing Necessary Truth, intues God."[3] I am not going, at present, to make any objection to this conclusion. All that I would remark is, a remark to be justified farther on,—if God is identical with necessary truth, what becomes of the notion of a *Creator?*

I now follow Mr. Ward in his continuation of this same question, in Section III. of the same Chapter, entitled " On the Relation between God and Moral Truth." He there takes up the thread of the argument at the point we have just reached, and draws a precisely similar conclusion with respect to the rela-

---

[1] On Nature and Grace, p. 43-44.
[2] Work cited, p. 44.   [3] Work cited, p. 45.

tion of Moral Truth to God. "This conclusion, that Moral Truth is identical with God, is based on the same grounds, which establish the parallel conclusion in the case of Mathematical Truth: in this case, as in that, such a conclusion is the only possible mode of avoiding objections, otherwise insuperable. But there are reasons in this case, of quite a different kind, which also press most strongly towards the same conclusion."[1]

The passage which I shall next quote, as throwing the final light on Mr. Ward's theory of necessary intuitive truth, occurs in an answer which he gives to an objection which he supposes on the part of an orthodox opponent. The objection is stated at considerable length,[2] and comes to this, that Mr. Ward's view detracts too much from the direct agency of God, and reduces it too much to a kind of pitiless destiny. Mr. Ward replies: "Now I say, secondly and more directly, that the objection is *not* efficacious at all against this doctrine. Let it be observed then, that the necessity of Moral Truth, as of Mathematical, is a *hypothetical* necessity. It is in no respect necessary that God shall create a triangle; it is only necessary that, *if* He create one, its three angles together equal two right angles. In like manner it is in no respect necessary, that He shall *create* beings possessed of freedom and intelligence; it is only necessary, that, if He *do* create such beings, they are subject to this or that moral obligation."[3] I need not continue the citation. I will continue it by analogy.—It is not necessary that God should pro-

---

[1] On Nature and Grace, p. 74.
[2] Work cited, p. 102-103.
[3] Work cited, p. 104.

duce at all: it is only necessary that, if He produces, He shall produce Himself; which self-production is His *Aseity*, whereby He is *Causa Sui*.

Observe the passing over into questions of *genesis* from questions of *nature*; from questions of τί ἐστιν to questions of πῶς παραγίνεται. Observe the separation of the existence, the *thatness*, of objects, from their laws, their *whatness*. Now either you can carry this separation up throughout, up to God, and then you get, as source of genesis a pure, arbitrary, undetermined *power*, without any analysis or nature at all, without any identification with necessary truth; or else you must make, from the first, a distinction only and not a separation of the two questions, and confess that the phenomenal world, in which the necessary truths are found, has a history coeval with its nature and analysis.

In other words, you cannot make truths at one time necessary, at another hypothetical; you cannot identify God with necessary truths, when you are asking *what* God is, and what His connection with them; and then, when you come to ask how God *acts*, sever those same necessary truths from their content (the triangles and the intelligent beings spoken of by Mr. Ward), and make their existence in the phenomenal world hypothetical. The existence of equality between the angles of a triangle and two right angles *involves* the existence of a triangle, just as much as *vice versa*. Both are existent objects of reflection in the same sense. If the former is a necessary truth, so is the latter. There is no more dependence on the power of a creator in the existence of a triangle than in the existence of this characteristic of its angles. Mr. Ward's point of view has changed, that is all. The objection has been brought, that the *agency* of God must be upheld. Accordingly the "Creator" doctrine is brought in, and the necessary truths, which God was previously said to gaze on eternally, He is now said (by implication) to create by creating the mathematical figures, and the intelligent beings, of which they are the laws. And this, notwithstanding that necessary truths are thus separated from phenomena, in which alone they have meaning, and become χωριστὰ εἴδη παρὰ τὰ πράγματα, in the teeth of Aristotle and even of St. Thomas Aquinas. "Plato, thou reasonest well!"

§ 7. Here I conclude the examination which seemed to me requisite of the writings of these two distinguished men, the one, Father Kleutgen, a conceptualist, the other, Mr. Ward, an intuitionist. Both theories we have found to lead to the same desired haven, the conception of a Creator. Both theories come to this conclusion because both start from the same fallacious principle, the separation of object and subject, by adopting the method of direct perception in philosophy. What is violently sundered must be violently reunited.

The catenation of thought is clear. As soon as any one (1) starts with object and subject separated, and then (2) enquires into the *history* of things, he is landed in the conception of a *cause*, of a cause of that cause, and so on without assignable limit. Mere weariness compels him to look about for some halt. He says,—How long am I to go on with this search for prior causes? The answer is plain,—Until he has accounted for all the phenomena which he knows or suspects to exist in the world. At that point and not till then he can stop. Oh, he says, that is soon done; I cannot indeed imagine all the phenomena which I suspect to exist, but I can do more, I can imagine that I imagine them, and more, indeed all that are *possible*, whether I actually suspect them or not. And I will imagine my Cause endowed with a fund of force sufficiently great to produce all these effects. This will then be the cause of everything that is possible; cannot therefore be caused by anything; but will be the first or uncaused cause of all things.

The conception of First Cause is thus easily reached, but it is not so easy to clear it of contradictions. It is a *particular* thing, by the very conditions of its origin, from separation of object and subject in direct perception. Nevertheless, there was evidently a time when it was itself *the All*, namely, before it created the world. The conception of the All, therefore, is not wholly incompatible with it. But when it has created the world, then, as a cause and not the All, it is a particular thing, and as such must have a cause. But again, as it is itself the cause of *everything else*, its cause can only be itself; it is Causa Sui. It is therefore *Causa Sui et Mundi;* and this, I believe, is the full scholastic conception.

But in the next place, if it is its own cause, it follows that it is also its own effect, *effectus sui;* and since cause and effect are correlative and coextensive, its effects as well as its cause are itself. For if A (as cause) contains anything for which A (as effect) has no correspondent, then A (as cause) is not cause *of itself*, but only of part of itself; and yet the sole *raison d'être* of A (as cause) is to be cause of A (as effect), that is, of itself. A (as effect) is that which is the basis of the whole reasoning. Consequently A (as cause) cannot be cause of itself and of the world too, *causa sui et mundi*, unless the world is part of itself, part of A (as effect).

Just as A has no cause but itself, so also it has no effect but itself; and we return again to the point from which we started, namely, that the First Cause is in some sense also the All ; only that now we find it to be so after the supposed creation of the world, as well as before it. In other words, the world and its cause are one.

There is, then, no resting in the notion of First Cause, or Causa Sui. The term *causa* imports particularity and separation of parts ; the whole term *causa sui* imports universality and union of parts. The term *causa sui*, as remarked in a previous Chapter,[1] is an hyperbole, caused by the attempt to conceptualise, or express in terms of *thought*, the infinite object of perception. When recognised as such an hyperbole, and consequently freed from any endeavour to draw consequences from it as if it were a statement of the truth of things, it remains comparatively harmless. But it is no longer harmless when understood, not as an inadequate expression for the All, but as a statement that there is, as a matter of fact, a separate Entity called the First Cause, on which other entities, called created things, depend. Its particularity and separability, the very features which render it self-contradictory, are those which prevent its being harmless. And what I mean by its

---

[1] Chapter VI. Vol. I. p. 408.

particularity and separability is this. The first cause, though inseparable from itself, as *causa sui*, is yet separable from the world, its consequence and effect; it is not the world over again, subjectively, but the *cause* of the world. It can exist without the world; but not the world without it.

This feature enables it to be used as the source of a Revelation, in the usual sense of the word; as the author of Miracles; as the dictater of Creeds; as the establisher and sustainer of Churches. Whether Churches are supposed infallible in their personal representatives from time to time existing, or as the guardians of an immutable and infallibly true creed, makes little theoretical difference. In either case the authority of the Creator is appealed to in support of foregone conclusions, by virtue of the conception of a particular maker and revealer.

What is still worse than the injury thus done to philosophy is the injury done to religion. Religion, of which Churches ought to be the guardians, is implicated with the various systems of dogma which grow out of and around this erroneous notion of a particular Creator. And then even the healthy and necessary reaction against this narrow and false conception, in its opposition to Churches which insist upon its being held, and are at the same time the recognised guardians of religion, injures religion. Churches and the dogma of a particular Creator to which they cleave are the common enemy of philosophy and religion. It is the separation of the Creator from his Creation, a separation which renders him a particular and finite being, proposed to us by the churches as an absolute and infinite one, that it is essential to avoid. There is a sense in which the term *creator* might still be logically and profitably used. And the limiting conditions under which alone it can receive this sense, conditions which will modify the notion itself as now received, will be pointed out in the following Chapter.

It is not the facts in human nature, it is not the facts of perception, of knowledge, of feeling, of conscience, of moral sense, or sense of right and wrong, of love to man and love to God as an Ideal Object, that any true philosopher wishes to disprove, or banish, or weaken, or oppose. It is not Religion that is in danger from philosophy; it is the false dogma with which religion is associated. It is from its bad companionship with untenable dogmas, and with societies which make it their business to maintain these dogmas at all costs, that religion is endangered. It is endangered by the moral and intellectual anarchy, which the resolution to support a foregone conclusion will not let die.

If it is true that the fortunes of mankind, and even its continued existence, are bound up with the maintenance of true morality in life and conduct, and that the maintenance of morality again largely depends upon its being enforced and vivified by religion, it becomes an imperative duty at all hazards to tear religion away from a dogma which sets it at variance with the free exercise of thought, and weakens the authority of its sanction by undermining its claims upon the intellect. At the same time, the other and positive side of the same task, the negative side being provided for, becomes more than ever important; namely, to exhibit religion as not torn away from, but more closely than ever leaning on, its true supports in the unseen and eternal world; and as, in this way, bringing to bear the whole weight of our conceptions regarding that world to uphold and sanction whatever conscience shall declare to be holy and just and good.

CHAPTER XI.

THE SEEN AND THE UNSEEN.

Sed tempus lustrare aliis Helicona choreis,
    Et campum Hæmonio jam dare tempus equo.
                                                    Propertius.

Book III.
Cii. XI.

§ 1.
Demarcation
of the Unseen
World.

§ 1. I HOPE my readers have not skipped the foregoing Chapters to pounce upon this, attracted by the easy theological look of its title. Without reading those, this cannot be understood; but that is their affair, not mine.

It was said at the beginning of the foregoing Chapter, that to ask if there was a constructive branch in philosophy was to ask if Time, Space, Feeling, the Postulates of Logic, and the Axiom of Uniformity, had universal and necessary validity. Now if any one of these five has such validity, it would be enough to constitute the constructive branch of philosophy; for we should be in possession of one piece of knowledge at any rate, concerning the unseen as well as the seen world. It is not how much we know about the unseen world, but whether we know anything at all about it, that is important in deciding the

Book III.
Ch. XI.

§ 1.
Demarcation
of the Unseen
World.

question whether there is a constructive branch of philosophy, and whether philosophy for that reason embraces a larger object than science.

Many people will say that such small knowledge as we can attain of the unseen world is really nothing, tantamount to nothing, that we might just as well call that world "unknowable" at once and have done with it. If for *really* nothing they would say *figuratively* nothing, or *comparatively* to exact science nothing, or for *popular declamatory* purposes nothing, they would be nearer the truth. But even then they would be far enough from it. For the difference is enormous. Without this "nothing" we should be handed over, bound hand and foot, for the priests and the materialists to fight over with interminable wranglings. With it, we can criticise both the priests who profess to know everything, and the materialists who profess that nothing can be known. We can show precisely what can be known and why, why so much and no more.

On the other hand it is equally unjustifiable to pitch upon any of the five, or indeed upon any feature whatever, which shall appear to have universal and necessary validity, as the key-note, so to speak, of the universe, as the principle at the root of all things, as the secret of the Absolute. To construct the universe out of such partial knowledge is impossible. We have, it is true, the knowledge that the feature or features in question enter into the constitution of the universe at large; but we have no knowledge as to what part they make, what function they perform, in it. To say that the universe is a development of Will alone, with Schopenhauer, or of Thought alone, with Hegel, is naive. Granted there is some-

BOOK III.
CH. XI.

§ 1.
Demarcation
of the Unseen
World.

thing that answers to will, and something that answers to thought, in the universe. Still we know too little of their modifications to be justified in considering these features as the source and end of all. We know something about the universe, but we have not the key to it; we have it not in the form of "The Absolute." Theories of this kind rest on the confusion which was pointed out in the first Chapter, between the analytical and the constructive branches of philosophy; inasmuch as they profess to construct the universe as a whole with the materials given, and in the shape in which they are given, by analytical philosophy or metaphysic.

But if the seen and the unseen worlds together constitute the universe, the question next arises, how are these two worlds demarcated from one another, and how can the unseen world be introduced as a reality, being unseen? It is in answering this question that the principle maintained in the present work as the cardinal principle of philosophy, namely, the principle of Reflection, must if anywhere prove its competence and its validity. And I think that it will do so. Reflection not only takes into its purview a far wider sweep of objects than direct perception does, but it supplies means of distinguishing the point where the narrower purview ends, and where the wider purview continues alone. It is as if, fixing one foot of a compass, we traced out first an inner and then an outer circle from the same centre, that centre being self-consciousness or the principle of reflection. The inner circle is the seen world, the outer is the universe, and the space between the unseen world.

The seen world is Kant's *mundus sensibilis et intel-*

Book III.
Ch. XI.
———
§ 1.
Demarcation
of the Unseen
World.

ligibilis.[1] It contains whatever is or may possibly become an object of *direct* perception and thought to beings constituted as men are. The unseen world contains all that is or may be an object of *reflective* perception and thought to beings constituted as men are, if at the same time this object is or may be an object of *direct* perception and thought to other, differently endowed, beings.

It is, then, the distinction between direct and reflective consciousness, established in Chapter II., which furnishes us with the line of demarcation between the seen and the unseen worlds. It will be remarked, that direct perception is not excluded from the unseen world *in toto;* it is only *our* powers of direct perception that are excluded; and this it is which constitutes that world an unseen one. To differently endowed beings it would be a seen world; but we know it only as an object of our reflective powers. And this difference it is, namely, between the beings who have direct perception of the seen and unseen worlds respectively, which gives us the further distinction between the three branches of knowledge which were sketched out in Chapter I. For, first,

Reflection determines the range of metaphysic or the analytical branch of philosophy; and this range embraces as portions of it both the seen and the unseen world.

Direct perception in the seen world determines the range of science.

Direct perception in the unseen world determines that of the constructive branch of philosophy.

[1] Kritik d. R.V. Von dem Grunde der Unterscheidung, &c. p. 238. Hartenstein, 1853. And see also above, Chapter III.

§ 2. But it is time to turn to our list of five, Time, Space, Feeling, the Postulates, the Axiom, and to see whether all or any of these must be held to enter as necessary and universal features into the constitution of the unseen world, from being necessarily involved and presupposed in reflective perception. If any of them should be found not to do so, still it will not follow that this one makes no part of the unseen world, but merely that we have no ground for concluding that it does so. We shall not be entitled to put it into our picture of the unseen world, but we shall also be unable to deny its presence there. We shall be unable to say whether, in respect of it, the seen and the unseen worlds are coextensive and conterminous or not. For it clearly belongs to the seen world, and there it is that we are originally conscious of it; the present question being, whether it belongs also to the unseen.

About two of the five there can be no doubt, namely, Time and Feeling. There can be no consciousness without them, and consequently no existence. And be it observed that by *feeling* is meant more than one single undistinguishable feeling. There must be a plurality of feelings, and, since these must occupy some duration, there must be a chain or sequence of different feelings. But nothing whatever is implied as to the kind or kinds of feeling which must belong to the sequence. All that is asserted is that, just as existence means presence in consciousness, so consciousness means sequence of feelings, or change of feeling in time. Both conclusions are arrived at in previous Chapters.

Reserving for the present the question of Space, I take next in order the two last of our five, the Postulates and the Axiom, which we have already seen to be opposite aspects of each other, and therefore inseparable.[1] The question concerning them is of a nature to put in a clear light the precise character of my present theory of philosophy, its character as a philosophy founded on Reflection.

For it might seem at first sight that the Postulates of Logic and the Axiom of Uniformity have not that universal and necessary validity which would alone warrant us in affirming them of the unseen world. Change of feeling in time is the subjective aspect of existence, as both together are contemplated by reflection. The two aspects are the two halves of the total object of reflection; and the subjective half is the analysis, in philosophy, of the objective. Mere existence, then, would seem to involve nothing more, in a necessary way, than time and feeling.

Again, this same result might be reached by another line of argument. Granting that *esse* means *percipi*, it might be urged that it does not therefore mean *concipi* too, but on the contrary excludes it. The Postulates and the Axiom are the basis of all thought, conception, and reasoning; but though these presuppose perception, perception by no means presupposes them. There may be a perceived world which is not also a conceived one; a world of feelings in time not reduced into any order of thought whatever, not capable of being so reduced. In short the unseen world may exist as Chaos. It is not essential to existence that it should be subject to the Postulates and the Axiom.

The answer to both arguments is the same, and this answer will bring to light, more clearly perhaps

[1] Above, Chap. VIII. p. 103. and Chap. IX. p. 144.

than anything that has preceded, the precise character of the philosophy of reflection. This answer is, that the process or function of reflection is that which is the basis of philosophy, and not any one or more of its objects or results. When in the first argument it is said, that change of feeling in time is the subjective aspect and analysis of existence, this is but one result of reflection; it gives us existence in its lowest terms, so to speak. A second exercise of reflection gives us its total object, namely, existence and its subjective aspect together. But this perception, a perception of the equivalence of the two aspects and of their inseparability, involves the Postulates and the Axiom.

To the second argument it may be replied, that it has been already shown at length, in Chapter II., that reflective perception presupposes conception or thought to have taken place in remoulding the data of primary consciousness. But if so, it involves the Postulates and the Axiom, which are the terms expressing the logical machinery and leverage of thought. To exclude these involves the exclusion of reflection. And the result of including Time and Feeling in the unseen world, while excluding the Postulates and the Axiom, would be to reduce it to a world of primary perception, of percepts which are indeed the *data* and furnish the content of all knowledge, but which become knowledge only by being moulded by the Postulates and the Axiom. To knowledge these are as essential as the primary percepts. And the question as to the unseen world is not whether it is a richer object than the primary percepts, for this it must be if it is an object of knowledge at all, but whether, as an object of reflective knowledge, it is not larger than any object of direct knowledge; whether, in short, the direct method of cognition is not a restriction introduced into the larger method of reflective. It is not the mere fact of subjectivity in an object that makes it an object of reflective and not of direct perception. Feelings as well as things may be direct objects. It is the stopping short in the process of reflection that turns the object, at which you stop short, into an object of direct instead of letting it remain one of reflective perception.

There remains but one of the five to examine, namely, Space; and here a great difference discloses itself. Space, unlike Time, is not inseparably involved with any kinds of feeling but two, visual and tactual sensation, taking the latter in the large sense in which it has been already used. Space again is not requisite in order to reflective perception; the Postulates and the Axiom may be applied to sequences of feelings in time without supposing visual or tactual sensation, or the muscular or nerve feelings combined with them, to make any part of the sequences. There is therefore no reason for including Space among the features which we cannot but attribute to the unseen world. That world, as we are compelled by reflection to picture it, and not as we may picture it for other reasons unforbidden by reflection, is not a world occupying space, is not a world which is either visible or tangible. It is not what is commonly called a material world. It is a world of feelings subject to uniform laws, eternal in duration, but having no extent and no position in space.

This circumstance alone shows the necessity of a

constructive branch of philosophy. For how are we to picture the relation between two worlds so unlike as this, a seen world occupying an infinite space, and an unseen world occupying not space but an infinite time? If we restricted ourselves to what have been shown to be the *necessary* features of the unseen world, our conception of the relation, in picturing it, must be something of this kind; namely, that for a certain period in the history of the unseen world, a world of matter and space appears; the unseen world becomes spatial, visible, and tangible, at a certain point in its career; and this period, during which a spatial world exists, is the period of existence of the seen world.

But if, on the other hand, we are led, by considerations belonging to constructive philosophy, to attribute space, visibility, tangibility, to the unseen world, which nothing in philosophy prevents us doing,—then we make this unseen world of space eternal, by identifying it with the unseen world of time, which is necessary in reflection; and we also reduce, or elevate, it to the rank of differing only in degree, instead of in kind, from the seen world. For, on this supposition, it is unseen only because the qualities of visibility and tangibility, which it possesses, are in degree either too intense or too feeble to be apprehended by human senses; not because those qualities are different in kind from those of the same name in the seen world. This is the sort of unseen world which is apparently intended by the authors of *The Unseen Universe*, cited in a previous Chapter.

Several objections will, I foresee, be made to the restriction of the necessary features of the unseen world to time, feeling, the postulates, and the axiom, excluding space. In the first place it does not harmonise with the doctrine that time is the form of subjective, space of objective, existence. But this doctrine is based on a total misconception of the distinction between the objective and the subjective. Existence is not, as this doctrine supposes, given to us ready distinguished into objective and subjective portions; of which one contains everything that is without us, *extra animam*, the other everything that is within us, *intra animam*. This is a psychological view of matters, a scientist's view, not a metaphysician's. Its ultimate distinction is a separation between the mind and things outside the mind. It is a distinction founded, not on reflection, but on direct perception. The mind is assumed as an existent among existents outside it. But this assumption includes already the assumption of space. The doctrine, therefore, that space is the form of objective existence rests upon the application of the method of direct perception, instead of reflective, to philosophy, the falsity of which proceeding may now, I think, be taken as proved. Time and space do not differ at all in respect of objectivity and subjectivity, any more than, taken together, they differ in this respect from their own material element, feeling. All three are equally and alike subjective, equally and alike objective. For the distinction of subjectivity and objectivity is not found directly and immediately in nature, but is perceived and drawn in nature solely by reflective perception, as was shown at some length in the second Chapter.

Another objection will probably be drawn from notions which are current respecting the possibility of deducing time from space, or explaining it as a mode of space. It is true that space, from its greater number of dimensions, and from its being, as figured space, the object-matter of geometry, serves as a sort of logic of conceptions which have, as states of consciousness, an existence only in time. As for instance we speak of the line of sequence, the time-thread, in redintegration; and of the intension, comprehension, and extension, of concepts; and illustrate the relation of the terms in syllogisms by circles or squares. But this very circumstance shows also the greater simplicity and more fundamental function of the form of time. The very figures of space, which we use to interpret to ourselves the relations of feelings in time, require time for their own envisagement, and cannot be perceived without it. Simply as states of consciousness, without thinking of them as cases of space-consciousness, they have duration. Their peculiarity as parts of our space-consciousness is no explanation of their duration, the common and essential feature of consciousness generally.

Time is no modification of space; not, for instance, a fourth dimension of it. It may be possible, and I am far from saying it is not, *in the constructive branch of philosophy*, to imagine a fourth dimension of space, by means of imagining some of the attributes which it must have, or some of the functions which it must perform, if there is such a dimension in connection with the three known to us; as, for instance, that in it should be contained and preserved statically, and in infinite minuteness of division and subdivision, all movements of solid bodies which we now regard as taking place solely in time, as happening and evanescent. Something similar to this seems to be the conception of Delbœuf in his Logique Scientifique.[1] "Dans la mécanique, le temps étant une quantité mesurable comme l'espace, il s'en suit que cette science est en définitive une géométrie à quatre dimensions. La formule générale $f(x, y, z, t)$ $= o$ de toutes les fonctions mécaniques représente un solide. Le temps n'y figure pas comme *l'image mobile de l'immobile éternité*, mais comme une nouvelle dimension immobile ajoutée à l'espace."[2]

But it is one thing to say that motion is a fourth dimension of space, and another to say that duration is. Suppose the time, which we now take as essentially involved in motion, to be resolved into or interpreted as a fourth dimension of space, along with the motion itself,—still this space of four dimensions will have time beyond it, for it will have duration, and will exist without, instead of with, change. Not to have duration would be not to exist.

Lastly, the objection will probably be made, that it is impossible to conceive the unseen world existing without other features than those now enumerated as essential to it. If so, I would reply, this only shows more clearly than before the necessity of attempting at least to frame a doctrine of constructive philosophy. The enumeration of features which cannot be denied admission into any picture of the unseen world, an enumeration founded on reflection, is not intended to be an attempt at constructing that unseen world as it appears to the intelligences, if any there be, whose seen world it is. It is only the necessary analytical basis for such an attempt. It is not an

---

[1] Essai de Logique Scientifique. Liège, 1865, pp. 260 to 284.
[2] The same, p. 276.

BOOK III.
CH. XI.

§ 2.
Its necessary
characteristics.

enquiry into the history of the unseen world, it is only a partial analysis of its content.

It may possibly be true that, without solid matter existing in space of three (or even more) dimensions, no existence is possible, seen or unseen. We know it of the seen world, we may suppose it of the unseen. But it does not follow from the only conceptions of the unseen world which analytical philosophy entitles us to frame. Only what is necessarily involved in the function of reflection, which gives us the notion of existence itself, is necessarily included in the notion of existence beyond the limits where presentative experience of sense ceases. But few as are the features which are thus included, it would be a great mistake to suppose them unimportant. It remains now briefly to show, in what their importance consists, beyond that which they derive from the fact of their being a portion of the unseen world.

§ 3.
Basis of
Teleology.

§ 3. This importance lies chiefly if not wholly in the second pair of facts which we have found to be essential features of the universe, namely, in the Postulates of Logic and their opposite, and phenomenal, aspect the Axiom of Uniformity. And it consists in this, that the onward movement of consciousness in the line of time is at once and indivisibly a cognition and an action. That onward movement which in its abstraction is expressed by the Postulates, though entering into all concrete movements of thought and action, is a choice as well as a cognition; for it is only by attention, which is a mode of volition, that it takes place. We *fix* upon some feature in consciousness and distinguish it from everything else, and then follows its combination in consciousness with something which was by that first act distinguished from it. But it is this very volitional movement of consciousness which is a prerequisite of reflective perception, and which for that reason has been included among the essential features of the universe.

This essential feature of the universe is indivisibly as well action as cognition. Cognition is an act of consciousness. And every act of consciousness is a cognition. There is no separating the two. In their utmost simplicity, in the most abstract shape of either of them, both characters are involved. There is one onward movement of consciousness which is inseparably an act and a cognition. This double character, always distinguishable never separable, is what is important in the Postulates and the Axiom. Why it is so, and what are the consequences that make it so, we shall presently see. But first let us dwell a little on the fact itself.

It is usually held that there are three main and ultimate functions of the mind,—Feeling, Cognition, Conation; or, Feeling, Knowing, and Acting. I need not cite authorities; it is the current view. These three functions, it is held, cannot be resolved into one another; they are quite different, though often inextricably mixed; still, though we often cannot say where one ends and another begins, yet beyond this border land of uncertainty we find concrete cases which are clearly one and not another, feelings which are not cognitions or actions, actions which are not feelings or cognitions, cognitions which are not actions or feelings. Much in the same way as in physiology we come to organisms which are clearly animal and not vegetable, or vegetable and not animal, notwithstanding that, in the border land, we are often unable to distinguish animal organisms from vegetable.

Now the conception, or rather no-conception, which lies at the root of this view of the three functions of the mind, is a psychological conception and not a philosophical one. It is a survival of the doctrine of faculties of an immaterial substance. And it is an attempt to rationalise that myth in such a way as to adapt it for purposes of science, by taking away its contradictions to recently observed facts, without breaking with the conception of the mind as an immaterial substance. The faculties of the mind have been rationalised into functions of the mind; so many sorts of *operations*, classified as observation demands.

But if the mind is not an immaterial substance, this conception of function breaks down for want of a substratum or basis, of something which has or exercises the function, just as much as the old conception of faculty does, and for analogous reasons. The faculties broke down partly from untenability of the notion itself, partly also because they classified the phenomena in a way which did not accord with facts. The functions break down because there is no agent whose functions they are. It will not do to say that they are functions of the conscious organism. For the question is, how they are discriminated and classified,—by examining the phenomena of consciousness, or those of the organism? If it is replied, as it must be, that it is by examining the phenomena of consciousness, then it is manifest that they are primarily functions of consciousness and not of the organism. But the term *function* is no longer applicable. Consciousness is not a thing which has functions; it has states and modes, but it is not an agent which operates. It is true that, if the organism is the agent which has consciousness as its general function, then the various modes of consciousness may be secondarily called functions of the organism. But then the discrimination and classification of these modes of consciousness, or functions of the organism, must be committed solely to the analysis of consciousness as such, and not of the organism; which brings us to the second reason why the present classification of functions breaks down, its want of correspondence to the facts, as we shall presently see. We do not know how the organism functions so as to result in consciousness, and therefore we know still less how it functions so as to result in this mode of consciousness and in that.

I am not in any way arguing against this secondary attribution of functions to the organism. It is perfectly legitimate, when the functions have been discriminated by subjective analysis, to consider them as functions of the organism. But what I am arguing against is the allowing the notion of function to dominate over and interfere with our subjective analysis, which alone ought to ascertain what those functions are. And it certainly does interfere with it, when it brings three classes of operations, however broadly distinguished in practice for purposes of every day life, (feeling, knowing, and acting), up to the rank of three irreducible functions of the organism, or of the mind, or of consciousness, on the plea that, if we could follow them up into their origins, we should find them irreducible. The rough distinctions of psychology are thus allowed to override the

THE SEEN AND THE UNSEEN. 249

Book III.
Ch. XI.

§ 3.
Basis of
Teleology.

subtil ones of philosophy. The function of feeling is held to be feeling and nothing else; that of knowing to be knowing and nothing else; that of acting to be acting and nothing else. Psychology goes no farther than this, which may possibly, when supplemented with the admission of mingling and interdependence between the three functions, be enough for its purposes. But the separation of the functions is fatal to a true analysis of them. It stops analysis short by its assumption that the analysis into three irreducible functions is complete.

To this psychological, separatist, and scientist view of the three functions of feeling, knowing, and acting, I oppose the results of subjective analysis of consciousness, its states, its modes, and generally its phenomena, without mixing up with it the consideration of these being functions of the organism. And I find, taking these phenomena in their whole extent, first, that they all and each are distinguishable into two inseparable elements, form (time and space), and matter (feeling); secondly, that this analysis holds good for all, when taken statically or one by one; thirdly, that, when they are taken dynamically as well as statically, or when consciousness is considered as an onward movement of statical moments, the two elements, time and feeling, combine in a manner which is expressible by the Postulates of Logic; fourthly and consequently, that, while time and space are the form of percepts in their lowest terms, the phenomena being taken statically, the postulates of logic are the form of concepts, or phenomena taken statically and dynamically together. Lastly I find that this statico-dynamic movement, of which the postulates are the form, is, in virtue of one and the same feature, attention, (which subjectively is a *feeling*, and this is the root of the whole matter), both an action and a cognition. An act of consciousness means, in its lowest terms, a conscious state in which there is the feeling of attention. The attention is one feature of the act, and as such is what it is known as, that is, a special feeling.

We have then, in total result, one single chain of consciousness, in which feeling, action, and cognition, are inseparable elements, not separable functions, however closely they may be supposed to be interwoven after separation. The fact is that they are never separate at all. The distinction of form and matter is the dominant one; then one of the two kinds of form, time, gives us the distinction between static and dynamic modes of consciousness, and their union; and finally the form of this last mode is double, containing a particular *feeling*, attention, which makes it an act, and a distinction of time, capable of being marked by subject, copula, and predicate, which makes it a cognition. Action and cognition in short are a modification or particular combination of time and feeling. And the cognitive act is, in virtue of its double character of action and cognition at once, the common source from which flow the two streams, Practice, the science of which is Ethic, and Thought, the science of which is speculative or theoretical philosophy. This common source is the point at which Ethic springs from, or is united with, the main stream of consciousness.[1]

[1] I take this opportunity of replying to a criticism of Professor Bain, who has done me the honour of citing the classification of the emotions given in my *Theory of Practice*, in the third edition of *The Emotions and The Will*, *Appendix B*. "I do not see," he

This cognitive act, with its double character, is performed only by means of the postulates; and therefore we have in it the mould, so to speak, in which all the phenomena of the unseen world, as well as of the seen, must be cast. Let us now consider what the essential features of this mould are, what is included in it as a general fact, notwithstanding that countless things are excluded. In the first place, the distinction between better and worse, and therefore between pleasure and pain *in their utmost generality*, is included. Attention is expectance; we do not attend to anything, but in view of something which it would be better for us to know, or feel, or do, than not. Although there is no definite purpose or expectation, yet there is always the feeling that it is not indifferent whether we attend or not. This feeling of its not being indifferent to us implies and involves the distinction of better and worse. We cannot conceive existence at all without

---

says, "much that is gained by carrying the abstract and antithetic couple 'matter—form,' throughout the whole scheme. In many instances, I fail to discover the relevance of the distinction : hardly anywhere does it appear to me to have more than a superficial adaptation." I can only reply that, seeing matter and form are in my view the ultimate distinction in all states of consciousness when metaphysically analysed, I had no choice, in classifying the emotions, but to carry that distinction through the whole scheme, and to distinguish in every case the emotional character of the representations to be classified from the imagery which excites or is pervaded by the emotion, and which, by its time and space relations, gives a handle for thought to deal with. I still see no other distinction so necessary, so universal, as this ; and I think what is said above confirms the view. So that if I have failed, as I clearly have to Professor Bain, to commend it in application, I have no resource but to lay the blame upon my own defective powers of exposition.

conceiving it as capable of improvement, and that without assignable limit. For the necessity of so conceiving it lies in conception itself; it does not stop with any attained stage, but goes on so long as we continue to conceive existence.

Nothing is hereby implied as to the actual degree of goodness or badness, or of either the kind or degree of pleasure or pain, in existence, either seen or unseen. The expectation involved in acts of cognition may be fulfilled in ways worse than our wish as well as better, worse than our present state as well as better. All that is involved in the act of cognition is the fulfilment of expectation in one way or another; the inseparability of expectation from the act of cognition; and the inseparability of hope from expectation. *Dum spiro spero.* And this perennial character of hope is rooted in the deepest part of our nature, in the most elementary act of cognition. It is bound up with the conscious life itself; to cease hoping is to cease living.

Time is a form of perception; the principle of contradiction is the form of thought. The form of thought, then, it is which implies an expectation of the better. Whether the matter or content of thought is better or worse, this expectation of the better is involved in it. So long as reasoning continues there is effort and expectation. With a full and manifold content of thought, this becomes effort to make the good prevail, expectation that it will prevail. It becomes the formation of Ideals. But the form alone being necessary belongs to metaphysic; the specific content belongs to the constructive branch, being to us contingent.

While, then, there is no indication that the content of the unseen world is better or worse than the content of the seen world, compared to such beings as we actually are; while we have no grounds for affirming anything whatever about it in this respect; it is yet evident, I think, that it shares with this seen world the characteristic of being capable of improvement, that is, of giving room for hope of what is better than that, whatever it is, which is at any time actual. It is a world, like ours, in which there is endless potentiality both of good and of evil.

It is surely a most remarkable thing that the notion of good is bound up thus intimately with our conception of existence; that we should not be able even to conceive existence without conceiving it as becoming better; that to conceive existence is to conceive improvement. It may be said on the other side, that we cannot conceive a better without a worse; and that deterioration is therefore equally essential to our conception of existence. But though it is true that we cannot conceive the one without the other, yet, (and this is what I meant to point out as remarkable), the order in which the two, better and worse, come forward in conceiving existence is fixed and essential to the conception. It is an order from worse to better, and not reversely. In the act of cognition the worse is first, the *terminus a quo*, the better last, the *terminus ad quem;* the worse is at the beginning, the better at the end. This constitution of the cognitive act is the basis of Teleology. If this were not its constitution, there could be no such branch of knowledge, or at any rate it would be an accidental and precarious one.

Book III.
Ch. XI.

§ 4.
What is not
necessarily
included in the
Unseen World.

§ 4. Briefly to sum up our deductions concerning the unseen world, before proceeding to draw the final consequences from them, it may be said in the first place, that the unseen world is continuous with the seen, framed on the same lines, and having certain characteristics in common with it. The seen and the unseen worlds are two parts of the universe of existence, the extent of which is ascertained by these common characteristics. And these common characteristics are, that both worlds are phenomenal, consisting of phenomena occupying time, but not necessarily space; phenomena which are cognisable both as subject to uniform law and also as capable of change, under that law, from better to worse and from worse to better ; or, in other words, are the objects both of speculative cognition and of teleology.

Let us next take note of what is not included in these necessary characteristics of the unseen world. Personality is not included; and this is a most important point. I do not deny personality of that world, any more than I deny space of it; but I maintain that it, like space, is not a necessary and universal characteristic of existence; so that if we find it in the unseen world, it must be by considerations belonging not to metaphysic but to the constructive branch of philosophy. There is, it is true, a personality necessarily included in all existence, and belonging therefore strictly to metaphysic; but this personality is on the subjective side of the conception of existence, not on the objective, not in the unseen world which is a part of that objective aspect, not among its phenomena and characteristics. This personality is our own, that of the reflecting Subject

BOOK III.
CH. XI.

§ 4.
What is not
necessarily
included in the
Unseen World.

who perceives and thinks of the universe. It cannot without special reasons be carried over to the objective side of the conception, and attributed either to the seen or the unseen world.

This direct carrying over of the reflecting personality to the objective aspect of the universe is what Ontologists aim at doing, or profess to have done. The union of it, when so carried over, with the objective aspect of the universe is their Absolute. For instance Ferrier, having given a demonstration of the proposition, " All absolute existences are contingent *except one ;* in other words, there is One, but only one, Absolute Existence, which is strictly *necessary ;* and that existence is a supreme, and infinite, and everlasting Mind in synthesis with all things,"[1] proceeds to make some observations and explanations, among which we find this : " Thus the postulation of the Deity is not only permissible, it is unavoidable. Every mind thinks and *must* think of God (however little conscious it may be of the operation which it is performing), whenever it thinks of anything as lying beyond all human observation, or as subsisting in the absence or annihilation of all finite intelligences."[2]

Now this thinking of anything as lying beyond all human observation, as subsisting in the absence of all finite intelligences, is just what I mean by passing into the constructive branch of philosophy from the purely analytical. For it can be done only by abstracting, for the purposes of thought, from a fact which can never be actually absent, the presence of self-consciousness or reflection. The abstraction

[1] Institutes of Metaphysic, Sect. III. Prop. XI. p. 522. 2nd edition.

[2] The same, p. 524.

Book III.
Ch. XI.
——
§ 4.
What is not
necessarily
included in the
Unseen World.

is by way of an hypothesis which can never be actually realised so long as we reason at all, and which therefore does not really carry us beyond our own human reflection; for we can only make the logical abstraction of all finite intelligences by continuing to exercise our own.

There are really two processes involved, first, the generalisation of the nature of existence, that *esse* is *percipi*; secondly, the inference that there is a Divine Percipient. Ferrier would abolish the second, the inference, by simply substituting the Divine Percipient in place of our own subjective reflection; thus attributing to the former all the strict *necessity* of the latter. And this he is enabled to do only in consequence of his original confusion of *conditions* with *aspects* of existence, as shown in a former Chapter.[1] But in fact, our knowledge of *all* personality except our own is inferential. There is no difference in this respect between our knowledge of human persons and of divine; all alike must be known to us by inference founded on generalisation. And to make this inference in the case of the unseen world, that is, to infer the Divine Personality, is to pass into the constructive branch of philosophy.

So that when Ferrier says, "Each of us can unyoke the universe (so to speak) from himself; but he can do this only by yoking it on, in thought, to some other self,"[2] the expression is not strictly accurate. No such total "unyoking" is possible; and consequently the "yoking on to some other self" is not a substitution of one self for another, but it is an addition of another personality, an addition by infer-

---

[1] Chapter VII. p. 21.

[2] Institutes of Metaphysic, Sect. I. Prop. XIII. Obs. 2.

BOOK III.
CH. XI.

§ 4.
What is not
necessarily
included in the
Unseen World.

ence. The divine personality, which is an object of our thought, can never have the same necessity in thought as if it were our own thinking, our own Subject. The *nature* of existence, namely, that *esse* is *percipi*, is fully satisfied by our own reflecting personality. It is only when we put the question of *genesis* or history,—How comes there to be a world independent of us, that the question of the divine personality is raised. We are then really abstracting from the question of nature, that is, abstracting from our own reflecting consciousness; and consequently the answer cannot be drawn from the analysis of nature, though it must be in harmony with it. True, we may *infer*, in Ferrier's words, a "Mind in synthesis with all things," *i.e.* an ideally perfect intelligence, to whom all infinity is present. But that Mind is no substitute for our own; each is objective to the other; and, though both are Subjects, there is infinite difference both of kind and degree between them. The Divine Personality, therefore, is not a strict necessity of thought, not an essential feature in metaphysical analysis, but is a problem which belongs to the constructive branch of philosophy.

Another thing not included among the essential characteristics of the unseen world is individual immortality, the continuance or the renewal of the conscious life of individuals after death. Like the first point mentioned, this point also is neither affirmed nor denied by metaphysic. To do either the one or the other is to enter on the constructive branch. Immortality is not affirmed by metaphysic because it is not necessarily included in reflection. The subjective aspect of the whole universe is sufficiently provided for by the individual's reflective conscious-

Book III.
Ch XI

§ 4
What is not
necessarily
included is the
Unseen World.

ness here and now. It is not necessary to imagine this reflective consciousness itself existing objectively with an extent and duration equal to those of its objective aspect, the universe. To do so would be to commit the mistake, explained in a previous Chapter,[1] of making the subjective aspect of things into a condition of them, a *conditio existendi*. Our knowledge of the world is in no wise a condition of its existing, but only of its definition, of *what* it is known as being. And for this a single moment of reflection suffices, without our having to suppose it continuing coextensively with the things known. Just as the sight of a star in space exists in the eye and not in the star, so also our reflective knowledge of the universe in time exists in us now, and not in the universe which we reflect on as eternal.

On the other hand, metaphysic has nothing to say against the conception of immortality. For it is no argument against it, that the notion of a soul as an immaterial substance is exploded. We use the phrase indeed, immortality of the soul, as if the soul was an entity, and the only one capable of immortal life. But we can imagine many ways in which consciousness may be continued or renewed after death, without the supposition of an immaterial substance. At any rate we know far too little of the nature of the unseen world, to deny the possibility of such a continuance or renewal. What we do know is, that this seen world of ours is bound up with that unseen one, and that the connection of the two is as much hidden from us as the unseen world itself.

[1] Chapter VII.

BOOK III.
CH. XI.

§ 5
Problems and
Methods of the
Constructive
Branch.

§ 5. The two points now mentioned, a divine personality of or in the unseen world, and a future life in the unseen world for beings which belong to the seen world,—two out of Kant's three great questions of philosophy,—are problems, and to us the most important problems, not of philosophy at large, but of its constructive branch. They are problems which cannot be solved unless we can construct to a considerable extent the unseen world, that world of which we have no direct perception, but which we have reason to imagine as surrounding our seen world, enclosing it like a sort of Happy Valley of Rasselas.

Now this world may be unseen for either or both of two sorts of reasons; either, first, because our kinds of feeling are limited in number, or secondly, because the modes of form, in which those feelings are cast, are limited in number or incomplete. And the question is,—since our seen world is a portion of, and in some respects continuous with, that larger whole,—are there or are there not any facts in this seen portion which enable us to complete, or warrant us in completing, in thought, the continuity of the content, which is broken by these two sorts of limitation?

Let us see, then, what methods are open to us, and under what limiting conditions, for making out anything concerning the unseen world, in the direction indicated by the two problems we have mentioned. Without attempting to construct that world, let us see what possible approaches, more or less promising, there may be towards doing so. There have been many attempts at actual construction. But, as was shown in Chapter I., these have mostly been

Book III.
Ch. XI.

§ 5.
Problems and
Methods of the
Constructive
Branch.

attempts to construct at once the seen as well as the unseen world, by basing both constructions on principles belonging to analytical philosophy; so that the result appeared as an *a priori* system of absolute existence.

The distinction between science, analytic philosophy, and constructive philosophy, has not hitherto been sufficiently insisted on, nor have the relations between them been accurately enough assigned. But I find the distinction itself drawn, or at least adumbrated, in Isaac Taylor's *Physical Theory of Another Life*, and there serving as part of the basis upon which that theory is erected, a theory which keeps within the limits of the Constructive Branch. Having there spoken of "that sort of mental philosophy which turns upon the adjustment and exact expression of abstract notions, which is properly termed Metaphysics," he proceeds: "But we look wider when we think of intellectual science, and think of it as a branch of physiology. Thus understood, it not merely embraces more objects, but comes under methods of investigation that are more diversified. Metaphysics is analytic simply; but Intellectual Philosophy, while it employs analysis, rests mainly upon induction (in the physical sense of the term) and must employ as well hypothesis as observation and experiment."[1] Taylor gives no clear distinction of "intellectual philosophy" from *psychology*; what we have in this passage is the clear discrimination of a certain "supernal" and conjectural part of it from *metaphysic*. That is already a great step; but the distinction cannot be said to be fully drawn, unless

[1] *Physical Theory of Another Life.* By Isaac Taylor. Chap. XX. p. 329-330. edit. 1858. The first edition was in 1836.

BOOK III.
CH. XI.

§ 5.
Problems and
Methods of the
Constructive
Branch.

the *three* members are at once exhibited and contra-distinguished; the distinction of any two is imperfect without the distinction of both from the third.

What, then, is the bridge by which we must pass, if at all, to the construction of the unseen world? Or rather, since we know that the framework of that bridge consists of the postulates and the axiom, what is it that forms the solid footway upon it? This is and must evidently be an hypothesis; it is no other than that which is employed in physical science, the hypothesis of the *material continuity* of phenomena. It has been shown in a previous Chapter that, while the axiom of uniformity was irreversible and admitted no exception, it yet afforded no ground for prediction of any *content* of thought.[1] Nevertheless we do expect, and all science is built on the expectation, that no violent change, no marked difference, in phenomena, will occur, unless prepared and worked up to by gradual changes which, when discovered, are said to explain the sudden ones; as, for instance, in landslips or explosions. The law of material continuity, or continuity of content, *natura non facit saltum, non patitur hiatum*, is an assumption upon which we found every kind of reasoning about phenomena. This is no metaphysical principle or law, like the postulates and the axiom. Still it is indispensable in all empirical science. It is equally indispensable, and its use equally justifiable, in the constructive branch of philosophy. It is not a principle which science demonstrates or can demonstrate. Therefore science cannot limit its range of application. True, the demonstrations of science confirm our confidence in it; but then its applicability is increased

---

[1] Chapter IX. p. 152.

Book III.
Ch. XI.

§ 5.
Probatas as a
Method of the
Constructive
Branch.

with every confirmation. As a principle of reasoning it is a principle of expectation. And its range extends wherever reasoning, wherever the postulates and the axiom of uniformity, extend. But its certainty is a different affair.

What I now call *material continuity* is not the same thing as continuity in mathematic or in mechanic, which is practically the same conception as infinite divisibility of space and motion, and is I believe usually regarded as axiomatic in science. But it is a continuity between *kinds*, a continuity in classification, where a gradual transition is assumed to exist between things which seem to us *heterogeneous*, a transition which would be, if known, the explanation of the apparent heterogeneity. Such apparent heterogeneity exists in several instances in the known world, as I remarked in the *Theory of Practice*.[1] And the effort of science always is to reduce such apparent heterogeneity to sameness; which is assuming that there is material continuity at its basis. In short I mean by material continuity very much what the authors of the *Unseen Universe* mean by their Law of Continuity, namely, "that the whole universe is of a piece; that it is something which an intelligent being is capable of understanding, not completely nor all at once, but better and better the more he studies it."[2]

But there is this difference between my conception and theirs, namely, that I apply this assumption to an unseen universe which is not *physically* material, material in the sense of consisting of *solid* matter. All solid matter, visible and tangible in kind,

[1] § 49. par. 4 Vol. I. p. 340.
[2] The Unseen Universe, § 261. p. 269. 6th edit.

Book III.
Ch. XI.

§ 5.
Problems and
Methods of the
Constructive
Branch.

even the luminiferous ether itself, belongs to what I call the *seen* world. And still I maintain that the seen and unseen worlds are *of a piece*, notwithstanding that we cannot find objects of direct perception in the latter. In a world supposed unseen, we can imagine objects proper, objects of direct perception, only where we can apply the notion of causation by motion of solids. The very difference of the two worlds arises from this, that the motion of solids fails us. And the distinguishing characteristic of the constructive branch of philosophy, from science, is, that we have only elements of nature or analysis of phenomena to build with, and have to assume a composition of them which is *not* the causation known to us.

Now this assumption is enabled and justified by the dissolution of the ontological or absolute character in physical causation or Force. If there was anything specific, any real or final explanation in *force*, as exhibited in the motion, or action and reaction, of solids,—then its failure or absence could not be supplied by an assumption of material continuity, from which this essential element, action and reaction of solids, was dropped out. But if, as is really the case, the action and reaction of solids is no more than *a case* of constant connection of antecedent and consequent in time; is only a form in which inseparable connection between phenomena appears to us, in direct perception, and not the ultimate account or explanation of that connection; then an inseparable connection, a continuity of apparently discontinuous phenomena, can be equally well assumed in a world of non-solid as of solid matter, that is, in a world where there is no *physical* matter and no *physical* force.

Book III.
Ch. XI.

§ 5.
Problems and
Methods of the
Constructive
Branch.

What we have to do, then, on this assumption is, to see what are the general possibilities of the unseen world, suggested by possible *completions* of phenomena which we find in the seen world. Taking the two sorts of reasons for either or both of which, it has just been said,[1] the unseen world may be unseen, we shall find that they break up into three heads, when the inseparability of matter and form is considered. There is, first, a possible incompletion of modes of form considered independently; secondly, a possible incompletion of modes of feeling considered independently; and thirdly, a possible incompletion of modes of feeling and form together, that is, (since we shall find ourselves restricted practically to *sensation*), of empirical physical existence.

I shall take the first and third of these heads first, postponing the second for convenience of exposition; since it is mainly on speculations falling under these two heads, which include questions concerning the connection of consciousness itself with physical matter, that our first Problem depends, namely, the possibility of a Future Life. That is to say, its possibility depends on the general nature of the unseen world; and our knowledge of this depends on our being able so to complete the mathematical, physical, and physiological sciences, as to reproduce the whole picture to which they belong from the square of canvas detached from it by our human modes of perception; or, to take another illustration, so to unmask a veiled face as to restore its true meaning and expression to a peering eyeball.

I. Under the head of incompletion of Form comes the question whether other modes of form besides

[1] Above, p. 259.

Time and Space may not exist in the unseen world; but this is a question to which we have no means whatever of even imagining an answer. We cannot imagine a third member in a series of which time and space are the two first members. Secondly, there is a similar question relating to a second dimension of Time. This too is incapable of an answer. The one dimension of time which we know, while embracing everything that exists, gives no clue to construct in thought a second dimension. But the case is different with the third question which may be put, namely, whether there may not be a fourth dimension of Space. Here we have already a series of three dimensions to aid the imagination; and accordingly we find that a fourth dimension of space, by analogy with the rest, although not capable of being positively pictured to our present faculties, is a favourite imagination with several mathematicians and physicists. And this clearly carries us over into an unseen world.

Book III.
Ch. XI.

§ 5.
Problems and
Methods of the
Constructive
Branch.

II. Under the third head, which I take next for convenience of exposition, comes the question of the origin of physical Matter in space. Physical matter, in its metaphysical analysis, is a combination of sensations of sight, touch, and muscular and nerve sensations, which give us the notions of solidity and resistance. The question as to the *mode of origin* of this combination, taken objectively, is one which can be put, but to which no answer can be imagined, in *science*. This is because all our notions of causation, in science, presuppose matter already formed and active. At the same time the question is not absurd, because physical science itself has long got rid of the old notion of Force as an efficient entity explanatory

Book III.
Ch. XI.

§ 5.
Problems and
Methods of the
Constructive
Branch.

of causation; and recognises, with some metaphysicians, that the *sum of the conditions* is the true meaning of causation. The question, too, is independent of that relating to the infinity of space; for there is no reason, but much the contrary, to suppose that physical matter is coextensive with space. The question then is simply this: What combinations of states of consciousness, if any, are the *condition* preceding or accompanying that particular combination known as physical matter? It does not seem possible that any light on this question should ever be within our reach.

Next to this comes a group of questions relating to the connection of consciousness with physical matter. These do or do not carry us up into the unseen world, according as they do or do not involve modes of consciousness new to us in point of kind; and therefore depend on questions belonging to our third head. Of the former sort is Taylor's " Second Conjecture," in the work above mentioned, " the abstract probability of the existence, on all sides of us, of an invisible element, sustaining its own species of beings;" in fact, that, " within the space encircled by the sidereal revolutions, there exists and moves a second universe, not less real than the one we are at present conversant with," &c [1] And this supposition has the closest analogy with one of Fechner's, which runs through most of his writings, and is expressed with great clearness in one of the latest, namely, that a subtil ether is one means of communication between all parts and all beings of the universe, and is at the same time the vehicle of the consciousness of a single universal being, comprehending the whole;

[1] Physical Theory of Another Life, p. 239–240. edit. 1858.

Book III.
Ch. XI.
—
§ 5.
Problems and
Methods of the
Constructive
Branch.

so that we may regard a beam of light as a nerve stripped of its albuminous sheath, and the universe as traversed in all directions by such nerves.[1]

Such are some of the problems which present themselves when we attempt to construct in any direction a picture of the unseen world. I confine myself to mentioning, and make no attempt to criticise, still less to offer any suggestion of my own. The general problem consists of three terms, the seen physical world, our consciousness, and the unseen world. What are the relations, first, between any two of these, secondly, between any two and the third? That is the problem.

I shall perhaps put my meaning in the clearest light by devoting a few words to a work which I have already had occasion to criticise, Dr. MacVicar's *Sketch of a Philosophy.*[2] His system is one of the old or pre-Kantian sort. He divides existents into three classes, spiritual, ætherial, and material.[3] And he constructs the worlds of physics and psychology by supposing that God, the eternal Spirit, " awards existence" to ætherial and material things and to other spiritual beings, in accordance with one supreme law, which he names the Law of Assimilation, and its derivatives to the number of six, which fall into two sets. owing to "the twofold fact that the finite assimilates itself on the one hand to the Infinite, and on the other hand to itself."[4] Gravitation, inertia, elasticity, electric, magnetic, and chemical action, and the chief

---

[1] Einige Ideen zur Schöpfungs- und Entwickelungsgeschichte der Organismen, p. 105-107.

[2] See back, Chapter VII. p. 28.

[3] Work cited, Part I. p. 34. Part II. p. xi.

[4] Work cited, Part II. p. ix.

Book III.
Ch. XI.

§ 5.
Problems and
Methods of the
Constructive
Branch.

functions of living organisms, nutrition, reproduction, and heredity, are deduced as cases of this one great law of Assimilation, in its two modes of operation, namely, analytic and synthetic action.[1]

The ætherial element is created to mirror the infinite *Being* of God, as the world of spirits is created to mirror his infinite *Power*.[2] And then from the æther, under the law of assimilation in its action already described, are produced first material organisms, and secondly, from these, inorganic matter.[3] Animal life is conceived as previous to vegetable life and as one of its conditions, just as organic matter in general is to inorganic.[4] Finally, the organic and material world culminates in the development of beings, man being foremost, whose nerve and brain systems are permeated by a homogeneous, hyaline, and invisible, substance, an æther, which is in its turn the abode of a spirit capable of communion with the Author of all.[5]

What strikes me as most admirable in this system, which I have only been able most imperfectly to sketch, is the truly philosophical application of the law of symmetry, as a case of the general law of assimilation, to construct the various sorts of material molecules which chemists find to be indecomposable in the laboratory, and which they are therefore con-

[1] Sketch of a Philosophy, Part I. p. 53-4. p. 113. Part II. p. xv. Part III. p. 11. and Chap. IV. Part IV. Chapters V. VIII. X. XI. And many other passages.

[2] Work cited, Part I. p. 120.

[3] Work cited, Part III. pp. 135. 158. Part IV. pp. 6. 12. 40. 66-8. 81. And other passages.

[4] Work cited, Part IV. pp. 20. 80. 84.

[5] Work cited, Part II. p. xvii. Part IV. Chap. XVI. And other passages.

Book III.
Ch. XI.

§ 5.
Problems and
Methods of the
Constructive
Branch.

tent to treat practically as if they were ultimate substances in nature.[1] The 63 or more chemical ultimates, as to us they appear, are doubtless, Dr. MacVicar urges, *products* of physical processes, products which have endured a fiercer ordeal than they can be subjected to in human laboratories; and therefore must be results of laws of structure and composition similar to those which we can detect by experiment, or can discern by the microscope. The material continuity of nature goes on, where our powers of discernment stop short.

But it is not with this part of Dr. MacVicar's work that I am concerned here. It is not with his theories regarding the seen world, whether ætherial or material, that I have to do; but with his conception, first, of the spiritual world, and secondly of its connection with the physical world. His philosophy includes my *constructive branch* and my *metaphysic*, as well as physical science. It belongs to the constructive branch, not in virtue of his development of matter out of æther, and spirit-homes out of organisms, but in virtue of his deducing æther with all its properties from the divine "award of existence" and institution of the law of assimilation. His philosophy has really two parts, first, a constructive branch *of philosophy*, secondly, a constructive *physical science*, based upon the first;—an *a priori* science based upon an *a priori* philosophy.

He bases the science of the seen world upon the philosophy of the unseen world;—that in brief is my objection to Dr. MacVicar's system. And my reason is this, that the philosophy of the unseen

[1] Sketch of a Philosophy, Part II. p. x. Part III. pp. 5. 45. 157. Part IV. p. 45.

BOOK III.
CH. XI.

§ 5.
Problems and
Methods of the
Constructive
Branch.

world, when truly conceived, can be no basis for, but only an outcome from, the science of the seen world. Give to that philosophy no more than what strictly belongs to it, and it affords no basis for science. To make it such a basis is to import into it conceptions drawn from science itself. How, it will be asked, does Dr. MacVicar do this? The answer is, By assuming that spiritual beings are existents which exert force. Spiritual beings belong nominally to the unseen, but really to the seen world, because supposed to exert force. Dr. MacVicar's conception of spirit, therefore, is double; a spirit is a heterogeneous being; partly physical and force-exerting, partly conscious and rational. It is man in the concrete, or unanalysed, over again; a conscious organism renamed. I do not think this in the least detracts from the *scientific* value of Dr. MacVicar's work, and I only wish I could convey to the reader the admiration which I feel for it and the pleasure with which I read it. Without pronouncing any opinion whatever on its scientific results, this at least I may say, that it produces, on a non-scientific reader at least, the impression of being the result of deep and genuine insight into nature. But the philosophical character of the work is another matter; and this it is which is seriously injured by the conception of spirit which I have mentioned. Much more must be done in the constructive branch of philosophy, strictly so called, before it can be shown how, or in what general way, the unseen world or any part of it is the sum of the conditions for the appearance of the seen world. We have, in short, yet to work up to the unseen world, and let alone, for the present at least, the dream of coming down from it with an explanation or deduction of the seen world in our hands. Neither can the sum of the conditions of the seen world be taken, by a rough and ready mode, to be a "magnified man," half consciousness half force; nor can philosophy itself be heterogeneous, half science half theology.

BOOK III.
CH. XI.

§ 5.
Problems and
Methods of the
Constructive
Branch.

For where, it must be asked, where does Dr. MacVicar get his conception of God, the great spirit, as an all-wise, all-good, all-powerful, Being? Surely it must be replied, from the moral nature of man, from the phenomena of the practical reason, the phenomena of ethic. God is to us the Ideal of the practical reason, of the moral nature of humanity. This is what the term *God* means to us. We do not know God immediately *as a creator*. Consequently we cannot treat his existence as explanatory to us of the existence of the kind of world which we see around us. God is one of the problems of the constructive branch of philosophy, of which branch one basis is science, the other metaphysic. Physics, consciousness, and the unseen world,—these are the three terms of the problem which lies before the future.

One writer there is who seems to me to have both clearly seen and firmly traced the conditions of this problem, James Hinton, too soon taken from us, when just reaching the full maturity of his powers. It is from the point of view of this problem that his writings, for the most part, must be supposed to speak. In his well known distinction between the Apparent, the Phenomenal, and the Real, the two first terms coincide with Kant's *mundus sensibilis et intelligibilis*, and the last with my *unseen world*. Not a world of *Things-in-themselves* or Unknowables, this latter, but phenomenal in that sense of the term in

Book III.
Ch. XI.

§ 5.
'Problems and
Methods of the
Constructive
Branch.

which it is opposed to noumenal; a world which contains the conditions of our seen world, of what is accessible to our powers of direct perception, being itself inaccessible to them, but open to our conjectures and evidenced by reflection. This world is that which is the chief object-matter of Hinton's works.

And here I will record (what death has rendered a melancholy pleasure) that, but for conversations with Hinton in the two or three last years of his life, when first I had the privilege of knowing him and comparing his ideas with my own, the present theory of the Unseen World and the Constructive Branch of philosophy would probably never have been framed. *His* insight it was which stimulated me to the humbler (though still necessary) task of analysis and formulation.

Take one passage from many to show what I mean.[1] It is from a supposed Dialogue between Author and Reader. "*Reader.* I clearly see your meaning: one thing acts upon us, and another is consciously present to our perception. The former you call the Fact, and assert that it is spiritual or active; the latter is the phenomenon, and it is physical or inert. The spiritual truly exists, the physical exists only as an appearance.[2] If man were in a truly living state—not defective in his being—he would have feelings correspondent to the truth: but

[1] Man and his Dwelling Place. Book IV. Dialogue II. p. 233. edit. 1872.

[2] "As the appearance is to the phenomenon, so is the phenomenon to the fact. From an appearance, by considering our phenomenal relations, we learn the phenomenon, or that which is true to thought: from the phenomenon, by considering man's actual relations, we learn the fact, or that which is the very truth of being." The same, p. 237-238.

inasmuch as he is defective, his feeling is wrong." But the method on which Hinton most relies to give us knowledge of the real world, the world of fact, the world unseen, is not a method falling under any of the foregoing heads; it is not of a physical but of an ethical character, belonging to a branch which has yet to be sketched, the remaining supposition of an incompletion in the kinds of feeling with which we are endowed.

§ 6. III. Under this remaining head, I remark in the first place, that the question of the material continuity of the two worlds, in respect of the kinds of feeling, soon appears to be an ethical and not a physical speculation. For the *emotions* are the only branch of feeling which are incomplete in such a way as to allow us to imagine their further development. We can no more imagine a sense or an emotion specifically *new*, than we can imagine a new form of perception besides space and time. And in point of development, the senses have their completion in emotion. For they not only furnish the framework, or imagery of objects, to which the emotions attach themselves, and which they seem to pervade; but, according to Mr. Spencer's profound and luminous theory, emotion itself is nothing else than the remains of countless sensations, combined with each other, transmitted and enforced by heredity.[1]

Here, then, we have a probable link of connection with the unseen world. There is no mode of consciousness more developed than the emotions; they

[1] Principles of Psychology, Part IV. Chap. 8. Part V. Chap. 6.

are the τέλος of the sensations; and we know of no other kind of feeling which can be regarded as τέλος in turn to them. At the same time we cannot imagine a specifically new emotion, any more than a new sensation. It is not in this way that they will furnish a clue to the unseen. It is as the head and front of a development, as phenomena of conscious action and practice, as object-matter of ethic and not of physic, that they will enable us, if at all, to divine anything of the unseen world. What are the true principles and standards of conscious action, the true aim of morals; and what sort of nature in the universe at large is implied by the fact that those aims and principles are the true ones? For they cannot be true in morals unless they are founded in the constitution of nature as a fact. If ethical truth can be made out, then we know, on that ground alone, something of the universe in which it is a truth.

The methods, then, which give entrance to the unseen world, practically reduce themselves to two, that of the physicist and that of the moralist. The first leads only to those structural characteristics of the unseen world, which are the *conditio existendi* of the seen; the second only to its moral characteristics, upon which those of the seen world depend for their practical meaning and significance. The conception of a supreme intelligence or personality is to be reached only by the second method; causation, or the conditions of existence, only by the first. The first method goes backwards in order of history, to place its conditions of existence at the beginning of things; the second goes forwards in order of history, to place its ideals at the end. And this contrast in the two directions is determined by the different nature of the two methods, one theoretic the other teleologic, combined with their both adopting the order of time as their common basis.

In order to combine the results, whatever they may prove to be, of the two methods, in order to combine the Ideal of the moralist with the efficient agency of the physicist, and thus to construct the unseen world of a piece with the seen, it would be necessary to put together, as it were, the two directions, from end to beginning and from beginning to end, and by this means take the whole period statically instead of dynamically. The unseen world would then be regarded not as a process but as a state, a single existent, including the seen world as a portion of itself, and with it forming one immovable universe, infinite in space, eternal in duration, an universe which to us now, who live only in the seen world, is necessarily an Ideal, that is, something which can be thought of only as the goal towards which both our knowledge and our practice move, never attaining its realisation.

The teleology of the emotions, then, is the chief clue which we must take to guide us in attempting to bridge the gulf between the seen and the unseen worlds. They are the most developed outcome of the whole chain of physical causation as we find it in the visible material world. They seem to be broken off, to be as it were the rough transverse surface of a column deserted by the builders, wishes which not only are never realised, but of which the real object is not known to the wisher. They seem to have their completion neither in themselves, nor yet, as the sensations have, in something not themselves.

Now if we yield to suggestions of this kind, and

put the question, for instance, whether the broken chain of development, of which they are the issue, can be supposed to find its continuation at all; or only in other beings different from ourselves and greater, as one kind in the animal creation stands to another in scale of dignity; or whether it is possible that the continuation should take place in a life which is physically the prolongation of our present life; or again if we ask what state of the unseen world would be indicated as final, if the highest moral emotions were to be fully realised either in ourselves or in other beings;—we should find that such questions as these, springing from teleology, would refer us back to the former physical method, as alone capable of disclosing the possibility or the conditions of realisation. Thus the problem of a future life belongs, as was remarked above, to that first branch of the question, while the problem of personality belongs to the present or ethical branch.

But now let us consider this second branch, both as to the sort of information it gives, and as to its degree of certainty. As we have seen, it is in the idealisation of the emotions, which are the object-matter of practical reasoning or Ethic, that we have the chief clue to the construction of the unseen world. For the physical and physiological causes, which have given us the emotions originally, are themselves continuous with conditions in the unseen world *a parte ante*, and the prolonged operation of these same causes will therefore, on the same assumption of continuity, carry on the emotions, which are their effects, nearer and nearer to the realisation of their ideal end, *a parte post*.

Practical reasoning itself, which consists in actual

THE SEEN AND THE UNSEEN. **277**

Book III.
Ch. XI.

§ 6.
The aid given
by Ethic.

choice between emotions, is by its very nature and definition teleologic, proceeding by ideals;[1] it imagines a desired *end*, and the desire makes that end a final cause of action. Every act of reasoning is volitional; but practical reasoning is distinguished from speculative by its aiming at possessing or effecting the *good*, as distinguished (though not separated) from the *true*. Practical reasoning, then, inevitably and instinctively carries us over, by its idealising method, from the seen world to the unseen as containing its idealisation. And all conscious beings are agents who help to make the unseen world what it is. Their volitions are evidences of its otherwise unknown nature.

True, we cannot idealise the imagery, as we can idealise its emotional content, which latter we do by desiring it in greater and greater intensity. The imagery which is its framework cannot take ideal shapes corresponding to this increasing intensity in its content, just because we have no direct perception of the unseen world. We have to be satisfied with vague outlines, described mostly by *negatives* of the known attributes of things in the seen world, such as incorporeal, immaterial, imperishable.

The imagery which embodies the reflective emotions consists in imagining those emotions to be felt and reciprocated by persons. When we idealise by intensifying those emotions, we introduce personality into the unseen world. This whole subject has been treated in the Theory of Practice, to which I must here be satisfied with referring.[2] Personality is the

---

[1] I would refer the reader on this point to the Theory of Practice, § 56. Vol. I. p. 401-416. Analysis of Practical Reasoning.

[2] Theory of Practice, Vol. I. §§ 22 to 48. And again §§ 72. 73.

image, or representational framework, in which reflective emotions are embodied, and which, being
reflective, they necessarily involve. But the only
way in which we can idealise personality, when intensifying its emotional soul or content, is on the one
hand to abstract as much as possible from all corporeal attributes, and on the other to retain in the
greatest strength all mental or spiritual attributes,
whether of intellect, volition, or emotion. In this
way it is that we imagine the ideal personality of
God.

That there are two inseparable elements in this
process, emotion and imagery, in religious matters,
explains at once both how it is that religion is in-
grained in the nature of man, and why we are entitled
to affirm so very little about its object. The existence of religion depends on certain emotions and the
idealising tendency of practical reasoning; our knowledge of its object depends on the representational
imagery at our command, supplied by the direct
perception of the seen world. It is owing to the
activity and fertility of the imagination that creeds,
which are theories of that imagery, have been multiplied. And thus we see explained, also by the same
double element, why it is that every advance in
intensity of real religious feeling has been accompanied with the demolition of creeds and dogmas.

There are some who would persuade us that this
ideal personality is a pure fiction, an unfounded
fancy, standing on much the same grounds as the admitted fiction of Things-in-themselves, and destined
to pass away from the human mind like a dream
when one awaketh. The results disclosed by the
analysis and philosophy of reflection are the answer

THE SEEN AND THE UNSEEN. 279

Book III.
Ch. XI.

§ 6.
The aid given
by Ethic.

which I would make to reasoners like these. The reflective and imaginative emotional element in consciousness is not the seat of passing fancies, but is that strain in human nature which is the most permanent and certain of development. It is with this element that the idea of the Personality of God is inseparably bound up. This element it is, and not the intellectual search for, or supposed proof of, a First Cause, which gives us at once Religion and its Object, the two being inseparably united.

This idealising emotion is the religion felt and expressed by Jesus Christ; not perhaps what is commonly called *Christianity*, for that means usually the religion of the Church; but the religion of Jesus Christ, meaning perfect and entire love to God, involving love to man as its corollary. Christ in fact revealed to us, from his own deeper insight and emotion, the Nature of God,—a personality in the unseen world which is the object of love with all the heart and all the mind.

God defined as First Cause would be a separable and therefore a finite being, a χωριστὸς θεός. Aristotle's arguments against the Platonic χωριστὰ εἴδη are all applicable against his own χωριστὴ οὐσία, of which a χωριστὸς θεός is a case. The existence of such a "separable" Being would be demonstrable, if at all, in science and not in philosophy. Now it is notorious that it is *not* demonstrated by science. But, supposing for a moment that it were so demonstrated, still the Being whose existence was demonstrated would not be God, but a Demiurge, or subordinately creative deity, created to create the world. And why not *God?* Because we could not love with the whole mind and heart a finite Being, however powerful or beneficent. The Demiurge, a χωριστὸς θεὸς, falls short of the true *definition* of God, which is, *Object of love with the whole heart and mind.* This is the sense in which it is true, and ever will be, that religion, faith and love, the "foolishness of God," overthrows the pride of reason and of intellect; namely, that "the world by wisdom knew not God," who nevertheless is the dominant power in human nature. The existence of God can neither be proved nor disproved by science; but by reflection it can be made evident as a fact bound up with consciousness itself.

And now, having shown in what way practical or ethical reasoning establishes certain facts concerning the unseen world, and in so doing lays the foundation of the constructive branch of philosophy, let us turn for a moment to the reaction which these facts exercise upon practice. Philosophy, by enforcing the consideration of the continuity between the seen and unseen worlds, is really bringing to bear upon practice the sanctions of eternity. Sanctions, I mean, in the jurist's sense of the term. In the presence of such a consideration we cannot but walk warily. We are part of a vast order, with which it behoves us to be in harmony. To live is no jesting matter, to live well is no easy one.

Theologians, it is true, have well nigh made the sanctions of eternity ridiculous in our eyes by adherence to their traditional creeds, their puerile conceptions of what these sanctions are, how they operate, and how they are to be met. They have well nigh rendered them odious by claiming to be their sole authoritative ministers and interpreters, as if they knew by special revelation what the unseen world was made of, and the chief mysteries of its construction. But it would be a fatal error to reject a profound and fruitful conception of philosophy, because it has been travestied by theology. Its deep religious significance, indeed, is the very thing which renders its travesty possible. For the popular mind, deeply feeling this significance, expresses it in its own imagery, and the imagery expressing it is a creed. The felt importance of the truths which they are attempts to express is that which gives creeds their vitality.

It is not difficult to see in what way the sanctions of eternity bear upon practice. The consequences which an action, done here, may have in the unseen world are the sanctions attached to it. And the possibility that we ourselves may be sharers in those consequences is the thought which, whenever it is admitted, most forcibly brings those sanctions to bear upon present practice. It does so, however, only by enforcing, with a solemnity drawn from the vastness of eternity, the judgments of conscience upon actions as they present themselves in the seen world. It does not alter but enforces those judgments. True, it clears the vision by heightening the importance of the object, and fixing the attention of the observer; but still the judgment is delivered by conscience, by whatever method it may be enlightened or instructed.

There is another mode in which the sanctions of eternity may operate upon practice in this world, and in this case without the supposition of individual immortality, namely, by the supposition of superior intelligences, or a supreme being, witnessing all that goes on in both worlds, the seen as well as the unseen. This is a direct enforcement of the dictates of conscience without any appeal to the future conse-

quences which an individual's actions may have for
himself. It is simply a magnifying and intensifying
of his own judgments upon his own acts. It is a de-
vice which he can employ for rendering that judgment more searching and less liable to deception; for
it supposes his own knowledge of the secret springs
of his own action rendered more perfect and pene-
trating, without any change in kind. If there is any
practical maxim which more than another deserves to
be placed at the head of all application of ethical
science, it is, not such a maxim as Kant expressed in
his famous formula 'So act as that your action shall
be valid for all reasoning beings,' but 'So act as if
your action in all its parts, all its motives, was wit-
nessed by beings of perfect intelligence.' But the
maxim is one of application of theory to practice, not
of theory itself. It cannot tell us *what* is right, it can
only intensify our perception of it; and this it does
by appealing to the sense of shame and the sense of
honour involved in imagining other witnesses and
those more intelligent than ourselves.

Kant, in my opinion, was fully justified in holding that the practical reason went farther and deeper
into the nature of things than the speculative, although
he was wrong in supposing that it connected us in
any way with a noumenal world of *Things-in-themselves.* It is into the unseen world that it carries us,
with this it connects us, not with the fictitious world
of Things-in-themselves. The moral and imaginative
emotions, the spiritual nature of man, his practical
reason and conscience, phenomena which belong to
the ethical branch of philosophy, these give us our
richest material for filling up the picture of the unseen world, our best basis for the constructive branch

of philosophy, the outline only of which is sketched by analysis and speculation.

I fully adopt, then, Kant's doctrine, that the practical reason is the sole ultimate ground of our knowledge of the nature, and therefore of the existence, of God; an agreement, however, which by no means involves agreeing with the whole of his theory of the moral law. But my difference lies in this, that I couple this doctrine, not with the assertion of unknowable existences, or *Things-in-themselves*, but with the conception of an unseen world which is in its nature phenomenal and therefore knowable. There are two sources of knowledge available for framing the fullest conception possible to us of the divine nature, one immediate and independent, the other inferential and subsidiary; the former being the moral nature of man, the latter his knowledge of the existence of an unseen world. God, being our highest Ideal, must fill the unseen and the seen alike.

We may sum up in one comprehensive word, *Will*, all that can be regarded as the leading shoot in the tree of existence, as known to us in the actual world. But what the Will as an *entity* is, we do not know; or rather, we know this, that in analysis it vanishes as an entity, just as Force vanishes as an entity, in the case of inorganic and inanimate things. It is an expression for *action* in conscious beings, as force is for action in unconscious ones. Action runs through the whole universe; and the Will, as represented by its analysis, *choice in moral matters*, connects man with the Spirit or Action of the Whole, which we may call God.

To put anything but the Spirit of the Whole in the place of God, to call, worship, and regard as God anything else but the Spirit of the Whole, is to lead man's effort and will into a false track, is to mistake a side shoot of the tree of existence for the main shoot, or rather it is to deflect the main shoot and dwarf it into a side one. This Comte's "Humanity," for instance, does. This Schopenhauer's "Will" does not. And therefore Schopenhauer is, on this point at least, the main point of ethical metaphysic, right and sound; notwithstanding his untenable identification of the will with absolute entity, a conception incompatible with the facts of reflection.

If we retain the word *Creator* to designate the Spirit of the Whole, it must be understood in a sense from which the notion of origination *ex nihilo* is excluded. It is coeval with the whole, of which it is the spirit. Neither is it itself the whole, but only its informing spirit. It may be conceived as standing to the universe in the same relation as the mind of a man stands to the man and his earthly history. As we say that a man has a mind or intelligence by which he orders his life, and which is the connecting thread of his actions, so we may imagine an analogous thread of consciousness and purpose to be the connecting thread of the universe. This would be "the Eternal that makes for righteousness," to use the expression of Mr. Matthew Arnold,[1] manifesting itself in phenomena which are the history of the seen world. And the action of this power would be both necessary and free, just as we have seen that human action is, according as we adopt the statical and

---

[1] In his profound and genuinely philosophical, though popularly written, work, *Literature and Dogma*, p. 81, 386, and passim. Also in its sequel, *God and The Bible*.

transverse, or the dynamical and forward-looking point of view.

Nor is the analogy rendered false in consequence of the human mind being no longer considered an immaterial entity. We are entitled to use the word *mind* to signify the conscious life of an individual man, notwithstanding that the theory which hypostasised it as an entity is abandoned. Similarly the conscious life of the universe may be called the spirit of the whole, without implying an entity either separate from the whole or prior to it in order of existence. It is a strain pervading all things, seen and unseen,—not a condition separable from its conditionates. This is the limitation which was spoken of in the foregoing Chapter,[1] as that under which the term Creator might legitimately be retained.

It must never be forgotten that, when we enter upon speculations like these, we are taking up a point of view quite different from that of pure reflection or subjective analysis. We are following up or completing the results of that analysis in the constructive branch of philosophy, and are treating the universe, so completed, as if it were an object of our own direct perception, just as a conscious organism is the object of psychology, or as a mineral or a gaseous substance is of physical science. It is only that part of the universe which constitutes our seen world which is an object of direct perception to us. The unseen world and the universe which embraces it are only to be hypothetically treated in the same manner.

Herein lies the great difference between the philosophy of reflection and those Monistic philosophies,

[1] Above, p. 231.

some of which have been criticised above, which find the nature of the Absolute, or of the Whole, to consist in some mode of union between Object and Subject, in whatever way this union may be imagined; whether as a Substance of attributes; a power of Will; of Thought; of Self-manifestation; and so on. True, they thereby give a definite solution to the enigma of existence, but they do so only by dwarfing the universe, the total object of reflection, to its *half*, by making it again an object of direct perception.

My philosophy, on the contrary, leaves the universe uncircumscribed except by reflection; surrounding its seen and definable portion with boundless fields of unseen existence. Man is no longer regarded, as in some of these systems he is, as the top and crown of existence; his consciousness as that in which the Existent itself first attains to consciousness of itself; but man and this seen world of his are a speck in the infinitude of unseen existence, with the laws of which it behoves them to be in harmony. A Monist might say, that this was no solution at all; and doubtless it is not one *of the sort* that he expects and demands. It does not gather up all into a single term, and say—there—*that* is the nature of existence; but it shows that no solution *of that sort* can be given, because existence is known *not* by direct but by reflective perception, *not* by any perceived characteristic, but by the fact that we perceive characteristics at all. It shows that they who give such solutions have mistaken the nature of the enigma, and for that reason are wrong in the nature of the solution to be expected. It places us, in fact, in an infinite, partially known, universe, instead of in a defined and therefore finite one.

This seems to me a better and truer kind of
Copernicanism than Kant's, being a legitimate development, in philosophy, of Copernicus' own conception, the true position of the earth in the seen world
of astronomy. As the earth to the solar system, and
the solar system to the starry heavens, so the seen to
the unseen world.

The function of reflection, too, has its analogue.
The observer's position must always be on the earth,
and that position a *central* one, although the earth is
not the centre of the heavens. The astronomical
observer is *subjectively* the centre of his observed
world, and cannot escape from this position. So also
is the position of self-consciousness, or reflection, in
philosophy; central, not to be escaped from. And
from this position the philosopher sees more than the
astronomer, who sweeps the visible world with his
telescope, and more than the geometer, who measures
space beyond the reach of sight. There is an existence which only reflection can perceive.

In pure ethic also the position of reflection is
central. Moral self-consciousness is what is known
by the name of Conscience. From this there is no
possibility of escaping. But the content and purport
of its dicta, from man to man, from race to race, are
variable, depending on varying conditions, according
as knowledge changes, and the objects of fear and
hope are variously analysed and estimated. The
content of the moral law must be continually changing; there must always be an "old law" to be de-
fied and abolished, as the condition of advance in
moral life; but the reflective power which constitutes moral law itself is unchanged and unchangeable; we can no more break loose from it than the

astronomer can quit the earth to plant his telescope
in the moon.

§ 7. If my reader will do me the favour of opening his Coleridge's *Biographia Literaria*, and reading
the passage[1] which begins: "For a very long time
indeed I could not reconcile personality with infinity;
and my head was with Spinoza, though my whole
heart remained with Paul and John," down to the
words "strange anomalies from strange and unfortu-
nate circumstances,"—he will find that the problem
there present to Coleridge is precisely that of which
a solution is offered in the present Chapter, the only
solution known to me. It is a problem of what I
have now called the Constructive Branch of Philosophy. And the solution is contained in the word
Reflection, because this it is which enables us to
assign a content, and that a *personal* content, to its
objects, yet without shutting them up into a finite
shape, which we cannot help doing if we employ the
direct method of thought and perception. Reflective
perception alone combines unlimitedness in the object with selfness, that is, personality.

Of course I am far from saying that this is a
solution which Coleridge himself would have accepted, or which could possibly have been reached
on his principles of analysis, which lead him to rest
in a distinction of *faculties* as an ultimate one. We
know enough besides of Coleridge's system to be
aware, that he was looking for a solution in the

[1] Vol. I. Chap. X. p. 196-199. ed. 1817. And Vol. I. Part II.
Chap. X. p. 205-208 in the annotated edition of 1847.

direction of Theosophy; he was a Theologian more than he was a Philosopher. But all this does not alter the fact, that Coleridge's method of doubting, his way of *posing* the problem, was the true way, being the most searching of any. This it was which dominated the whole train of my thought, in aiming at a solution of the great problems of philosophy. My *intellect*, he says, was with Spinoza, my *heart* with Paul and John; and it was the *heart* which made him a theologian. The point, then, which forced itself upon my attention was this: That which hinders philosophical theories, otherwise satisfactory, from being true, the respect in which they fail of truth, is, that they do not harmonise the phenomena presented by the heart, that is, the *emotions*, with the phenomena presented by the intellect. Thus there arises a discord in the theories, which is not in nature. The facts, then, must be re-examined with this discrepancy in view. This was the dominating principle of investigation which I learnt from Coleridge.

Here then I bring to a close the present work, and with it the series of which "Time and Space" and "The Theory of Practice" are the first and second members. I have endeavoured to show the necessity of philosophy as a distinct branch of knowledge, if we would systematise all kinds of knowledge, possible as well as actual, in a manner capable of excluding intellectual and moral anarchy. A *system* is necessary for this purpose because, the moment any possible kind of knowledge is left unplaced, its cultivators become a schismatic sect. And the system must be *philosophical*, because no merely scientific one can embrace all kinds of knowledge, leaving none unplaced.

The means by which I have endeavoured to show this have been by giving, in the first place, a definite picture of the relation borne by philosophy to the various sciences, and secondly, an articulated scheme of philosophy itself, its method, its content, its divisions, and their concatenation.

I take up no position hostile to any special science, but on the contrary maintain that the truths which are discovered in any science take rank at once among the *data* of philosophy, among the facts to which philosophy must conform. *Nec mihi res sed me rebus subjungere conor.*

I take up no position hostile to religion, nor yet to traditional creeds, except in cases where their partisans attempt to impose them as philosophical truths, without other than traditional support. Creed-holders attempt to make their creed the intellectual basis of religion; philosophy discovers that religion has a basis firmer and deeper than a creed.

To inveterate specialists in religion, bent on making their own creed into a philosophy, and to inveterate specialists in science, bent on having no philosophy at all, these pages are not addressed. They are addressed to those who are at least open to the conviction, that philosophy has a legitimate and necessary function of its own, a history which determines its present state, and problems which in the future it may expect to solve. They have accordingly attempted to determine its general functions, its present state, and its future problems, with greater accuracy; and in so doing to give greater definiteness to the conception of philosophy itself.

The central point of the whole system is the same with the central point in all knowledge,—the phenomenal fact of reflection. It is in consequence of this fact, which becomes its principle, that philosophy exists at all, distinct from science. The gradual emerging of this principle into clearer and clearer light, its being grasped with greater distinctness by successive philosophers, is the key to the history of philosophy. What distinguishes the present system is, that it grasps this principle more distinctly than ever before, applies it more systematically, perceives more clearly that it is the vital principle of philosophy. The history and the theory of philosophy, the theory and the application of it, are three things inseparably interdependent. This, too, is a consequence of the principle of reflection.

SUMMA

(ADDED IN PASSING THROUGH THE PRESS).

1. The death of Force as an entity is the birth of the Unseen World, as condition of the seen or material world.

2. *Before* Reflection, the *is* of judgment is the *is* of the Copula; in reflection first, and then in its derivatives, it becomes the *is* of Existence.

3. The Existence which is knowable by Science is the Vestibule, to which primary consciousness is the Porch, and unseen phenomena the Temple. It is reflection by which these two additions are made to ordinary Experience.

4. Scholastics say *first*, There is an all-perfect Being, and *then*, We can love him with all the heart and mind. I reverse this order. I say *first*, We can love with our whole heart and mind, and *then*, The object of this love is GOD.

INDEX.

298   INDEX.

THE END.